Co-phase Traction Power Supply with Railway Hybrid Power Quality Conditioner

Keng-Weng Lao · Man-Chung Wong
NingYi Dai

Co-phase Traction Power Supply with Railway Hybrid Power Quality Conditioner

Springer

Keng-Weng Lao
Department of Electrical and Computer Engineering
University of Macau
Macau
China

NingYi Dai
Department of Electrical and Computer Engineering
University of Macau
Macau
China

Man-Chung Wong
Department of Electrical and Computer Engineering
University of Macau
Macau
China

ISBN 978-981-13-0437-8 ISBN 978-981-13-0438-5 (eBook)
https://doi.org/10.1007/978-981-13-0438-5

Library of Congress Control Number: 2018940417

Printed on acid-free paper

This Springer imprint is published by the registered company Springer Nature Singapore Pte Ltd.
part of Springer Nature
The registered company address is: 152 Beach Road, #21-01/04 Gateway East, Singapore 189721, Singapore

This book is dedicated to my family, supervisors and friends. It would not have been completed without their support. I would also like to dedicate the book to all those who are doing research on railway power and power quality compensation.

Contents

Abbreviations

A.T.	Auto-Transformer
ac	Alternating Current
APF	Active Power Filter
B.T.	Boost-Transformer
dc	Direct Current
DVR	Dynamic Voltage Restorer
HAPF	Hybrid Active Power Filter
HPQC	Hybrid Power Quality Conditioner
PF	Power Factor
PWM	Pulse Width Modulation
RPC	Railway Power Quality Compensator
STATCOM	Static Synchronous Compensator
SVC	Static Var Compensator
THD	Total Harmonic Distortions
UPQC	Unified Power Quality Compensator

Chapter 1
Introduction

Abstract Transportation is of major concern nowadays for city and country development. Rapid growth in transportation demand leads to a worldwide trend of developing high-speed railway. Co-phase traction power supply structure eliminates neutral sections and locomotive speed limitations. It thus has high potential for application in high-speed railway. However, in order to relieve power quality problems, power quality compensators are installed in railway power. The newly developed capacitive-coupled railway hybrid power quality conditioner (Railway HPQC) requires less operation voltage and is more beneficial than the inductive-coupled railway power quality conditioner (RPC). The co-phase traction power supply system with Railway HPQC thus has multiple advantages of (a) elimination of neutral section quantities and locomotive limitations; (b) higher transformer utilization ratio and reliability; and (c) lower power quality compensator operation voltage. In this chapter, the overall background of railway traction power, including its structure and worldwide development, is introduced. The power quality problems in railway power supply and existing solution methods are discussed, especially on flexible alternating current transmission system (FACTS) compensation devices. There are also discussions on recent research on railway power compensation.

1.1 Overview and Introduction

Nowadays, railway transportation is especially important for city and country developments. Electrified railway is preferred due to its beneficial characteristics: cleaner, safer, and more efficient. In order to satisfy increasing transportation demand, different countries have developed various plans on constructing electrified high-speed railway. For instance, according to "Revising the Long and Mid-Term Plan of the China's Railway" (Adjusted at 2008) [1], in 2020, China railway would cover more than 120 thousand kilometers, which forms a clear, well-functioned structure with electrification ratio of over 60%, that can satisfy the growing transportation demand. Moreover, the Chinese government is also having

K.-W. Lao et al., *Co-phase Traction Power Supply with Railway Hybrid Power Quality Conditioner*, https://doi.org/10.1007/978-981-13-0438-5_1

"8-hour railway life cycle" and "Four Vertical and Four Horizontal High Speed Railway Network" plans. Chinese citizens can travel from any place of China to the capital city Beijing within 8 h. Besides, the Macau government has also planned to develop a Light Rail Transit (LRT) to enhance city integration between Guangdong, Hong Kong, and Macau [2, 3]. These all show the importance of electrified and high-speed railway in modern transportation.

Ever since the emergence of electrified railway, different power supply modes have been used. This is due to the fact that different countries have developed their own electrified railway in history. However, they all suffer from power quality problems [4–7]. Major power quality problems in traction power supply include system unbalance, reactive power, and harmonics. Rapid and time-varying unbalanced locomotive loadings can cause the presence of negative sequence current in the three-phase power grid, which may damage the power system and reduce device lifetime. The inductive nature of locomotives also inject significant amount of reactive power into the system and threaten the system stability. The presence of reactive power also indicates inefficient usage of the power supplied. At the same time, nonlinear locomotives loads also inject large amount of harmonic current into the power system, causing additional power loss and even severe damage. Moreover, emergence of new power electronics techniques in locomotives also makes harmonic problem a more serious concern in traction power supply.

Thanks to the effort from power engineers and power electronics development, various power quality compensators have been proposed and used for power quality conditioning in traction power supply [8–10]. Since traction power locomotive load is dynamic, time-varying, and nonlinear, it brings about various power quality problems simultaneously. Fast dynamic compensation is therefore required. Compared to traditional passive compensation, active compensation devices based on modern power electronics techniques can provide better dynamic responses and thus more satisfactory compensation performances. STATCOM or APF is a good example of them and has been widely applied for traction power compensations. However, active compensators are not yet widely adopted due to its high cost compared to traditional passive ones. With this concern, hybrid filter was thus proposed for dynamic compensation performance with lower cost. With cost reduction and better dynamic performance, hybrid filter has higher potential to become widely adopted compensators over conventional ones. More about power quality compensators can be found in later sections of this chapter.

Considering the development worldwide, it is essential that the traction power supply developed in the future is applicable and suitable for high-speed railway. So far, most high-speed railway locomotives are electrified with 27.5 kV alternating current (ac) power. However, isolation components (neutral sections) are present in traditional ac traction power supplies. Locomotives lose power and speed when they pass through these neutral sections. They are therefore not quite suitable for high-speed railway application in the future to satisfy increasing locomotive speed requirement. Co-phase traction power supply is one of the newly developed systems which benefits in the elimination of neutral sections and locomotive speed limitations. Moreover, it can help to solve the traction power quality problem by providing unified power quality compensation. Co-phase traction power, therefore, has high potential to be applied in high-speed railway. In fact, the world's first co-phase traction device has already put into trial operation at China Chengdu Kunming Meishan substation. Various operation results have already been reported and published (refer to Fig. 1.1). It is one of the important projects supported by the Chinese government. Nevertheless, the railway power quality conditioner (RPC) requires high operation voltage to provide power quality compensation. The high RPC operation voltage induces high device ratings and cost, which may limit co-phase traction power supply development.

The application of hybrid coupling structure in power quality to reduce operation voltage is quite well known. The idea can, therefore, be applied to provide power quality compensation using lower operation voltage. However, the technique cannot be directly applied since active power transfer is also involved. The hybrid-coupled structured co-phase power quality compensator investigated is termed as railway hybrid power quality conditioner (HPQC).

Based on the considerations above, a low-cost high-speed railway power supply is developed. It is composed of a system based on co-phase traction with a novel railway hybrid power quality conditioner (HPQC). The system is beneficial for (1) elimination of neutral sections and reduction of limitations in locomotive speed; (2) enhanced transformer utilization ratio and power supply reliability; and (3) providing comprehensive unified power quality compensation with low operation voltage and cost. Detailed system analysis and design considerations are investigated and explored. Simulation and experimental verifications are performed based on practical data obtained from traction substations. The content organization can also be found in the last section of this chapter.

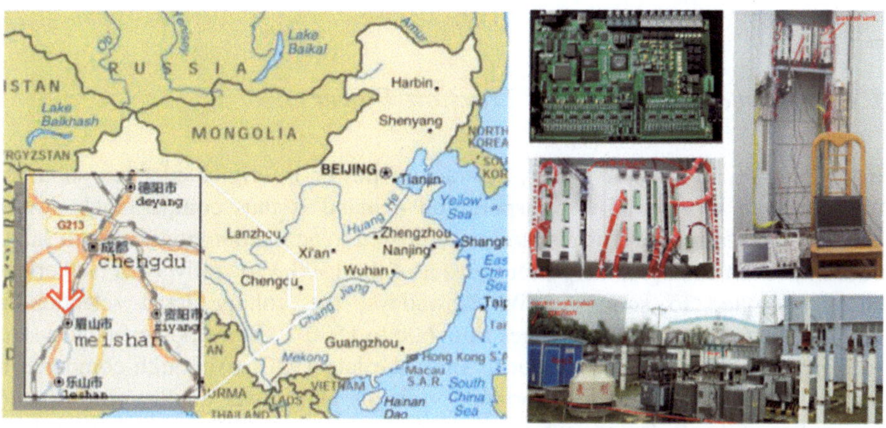

Fig. 1.1 The world's first co-phase device has already been put into trial operation at China Meishan substation. Shown in the figures are the control unit and appearances

1.2 Development of Electrified Railway Power Supply Mode

As discussed previously, different countries have developed different traction power supply modes, namely direct current (dc) and alternating current (ac) modes [11]. They are briefly introduced below.

1.2.1 Alternating Current (ac) Traction Power

ac traction power is mostly applied to long-distance high-speed railway. At the early stage of power development, ac power is preferred and many ac power devices are developed. The same case occurs in traction power. Originated since 1912, railway in countries like German, Austria, Swiss, Sweden, and Norway were electrified with 15 kV and frequency of 16 2/3 Hz. This setting is developed to minimize the interference of traction power with existing 50 Hz power distribution system. However, this usually involves installation of separate generation and distribution systems, which contributes to high initial cost. Traction power system running on 50 Hz was later desired. After the successful electrification trial of French State Railways with 50 Hz, 25 kV, similar system was widely spread over European countries such as Britain, Ireland, Portugal, Denmark, Finland, etc. Nowadays, ac electrification of 50 Hz, 25 kV is still widely adopted worldwide in long-distance railway, including China. Since ac electrification is normally used for long-distance railway, which normally requires high speed, ac electrification is therefore widely used in high-speed railways.

1.2.2 Direct Current (dc) Traction Power

In contrast to ac traction power, dc traction power is originated from suburban transportation services. The transportation demand is thus comparatively lower. Besides, the voltage level is also lower compared to ac traction power due to safety concern. As an example, the power transfer of the 1500 V dc railway in Netherland is only limited to some 5 MW. Nowadays, the voltage level selection in dc-electrified railway usually ranges from 600 to 1500 V. Some powerful dc system running at 3 kV was also introduced since the late 1920s. Although compared to long-distance railway the transportation variation and power quality problems are less severe in dc-electrified railway system, dc railway electrification is not suitable for long-distance high-speed railway.

1.3 Worldwide Development of Electrified Railway

1.3.1 China

As introduced, the Chinese government has already modified the plan of "Revising the Long and Mid-Term Plan of the China's Railway" in 2008 that by 2020 there would be over 60% railway electrification, with total railway coverage of more than 120 thousand kilometers. This is achieved so by constructing the "Four Vertical and Four Horizontal High Speed Railway Network" and implementing the "8-Hour high speed railway life cycle" plan. By doing so, Chinese citizens may travel from any part within China to Beijing within 8 h. It brings a lot of convenience and can push forward inter-city developments.

The high-speed railway development in China is quite or near to a world-leading role now. Besides the development, the Chinese government has also supported researches in traction power supply. The world's first co-phase traction is proposed by Universities in China and is supported by the Chinese government as one of the important projects. It therefore can foresee that the high-speed railway development will be under spotlight in China in coming years.

1.3.2 Japan

The high-speed railway development in Japan is also worth noticed. The world's first high-speed railway system, Shinkansen railway, was operated in Japan in around 1964, and is well known as the "bullet train". Its locomotive speed 210 km/hour was a leading high speed at that time. Until now, the development of high-speed railway in Japan is still gaining the world's attention.

Being the world's first high-speed railway operation country, Japan has putting much effort to further develop and enhance high-speed railway and to explore new techniques such as magnetic levitation vehicle.

1.3.3 America

Compared to other countries, the development of high-speed railway is relatively slow. So far, there is no standard high-speed railway in America due to area and geographical limitation. But the American government has also planned to invest and develop high-speed railway. The California railway is the first high-speed railway system in America. There is also other American high-speed railway construction plan. It is already stated by America President that their target is to over 80% of American people to use high-speed railway as traveling media by the year 2050. However, some American people are concerned with the high investment cost of developing high-speed railway in America.

1.3.4 Europe

The high-speed railway development in Europe is quite complicated since Europe is composed of many countries. Some typical examples are described below. For example, LGV, Europe's first high-speed railway system, was operated in France in 1981. In the future, France will also invest and develop high-speed railway techniques. The next example is Britain, which is known the world's first country to have railway transportation. Britain is having a plan to construct a major high-speed railway line connecting cities such as London, Birmingham, Leeds, etc. The whole plan is expected to finish around year 2033, and its final target is to shorten the traveling time from London to Manchester by around 45% (from 2 h 8 min to 1 h 8 min).

For reference, comparisons of different railway transportation power supply systems in other countries are shown in Table 1.1.

1.4 Traction Power Supplies

1.4.1 Various Traction Power Structures

Various traction power supply structures have been proposed and used in railway traction power. Two of them are discussed below, namely conventional traction power and new co-phase traction power.

Conventional Traction Power

Shown in Figs. 1.2 and 1.3 are the typical circuit structures of conventional traction power supply (Boost-Transformer, BT and Auto-Transformer, AT). In long-distance ac-electrified high-speed railway, three-phase power is usually transformed into outputs of two individual single phases through transformers located at substations. In conventional traction power supply, locomotive loads are connected across two single-phase outputs to obtain power from the source grid. Without proper action, phase mixing or short-circuit condition may occur since locomotives must run along the same contact wire.

Therefore, isolation components (neutral sections) are present at contact wire feeder lines between substations to isolate two single-phase outputs and avoid phase mixing. This, however, results in electrical isolation between various power regions. Power switching is thus required when locomotives pass through these neutral section isolation boundaries. Meanwhile, the locomotive is out of power. This causes locomotive power loss and limits locomotive velocity. Furthermore, excessive usage of neutral sections can result in undesired additional power loss. Conventional traction power supply structure is therefore not suitable for high-speed railway.

Table 1.1 Comparisons of different railway transportation power supply systems in other countries

	Germany	Japan	France	Italy		Spain
Name of the railway	Hamburg—Bremen	Shinkansen	TGV	Rome—Florence	Alta Velocita	Rome—Florence
Source voltage (kV)	110	154/220/275	225/275/400	130	130	220/132 (400/220 Madrid—Barcelona)
System short-circuit capacity (MVA)	>1000	>1700	9000	-	-	>2000
Power supply mode	ac 15 kV 16.7 Hz	ac 25 kV 50 Hz	ac 25 kV 50 Hz	dc 3 kV	ac 25 kV 50 Hz	ac 25 kV/ 50 Hz dc 3 kV at terminal ends
Power supply	Direct supply and BT feeding/ co-phase	±25 kV AT power supply, AT at every 10 km, when exceeding 30 km, a substation is located	±25 kV AT power supply AT at every 15 km	Supply with return wire	±25 kV AT power supply	Direct supply and BT feeding
Transformer connection and capacity	Single-phase transformer 2 × 15 MVA	Scott and WB transformer 60–200 MVA	Single-phase transformer 40–60 MVA	2 × 5.4 MV A rectifier	Single-phase transformer 40–60 MVA	Single-phase transformer 2 × 20 MVA
Distance between substations (km)	30–80	40–60	40–90	16	50	MAX50

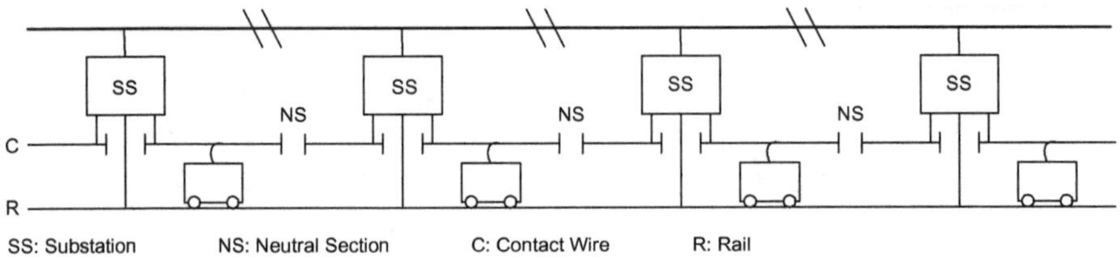

Fig. 1.2 Circuit diagram of conventional traction power supply (BT)

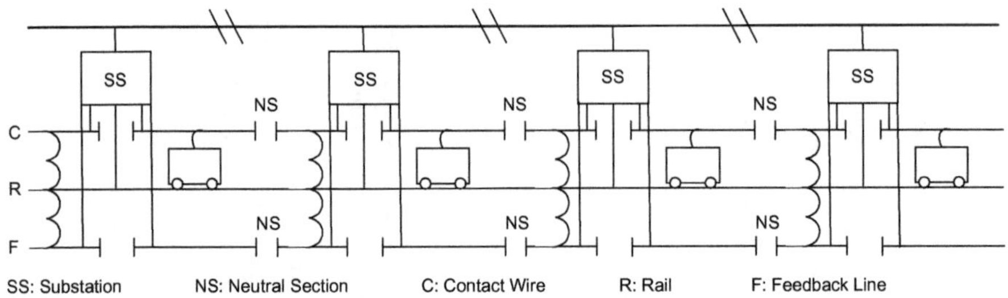

Fig. 1.3 Circuit diagram of conventional traction power supply (AT)

Co-phase Traction Power

Co-phase traction power supply is one of the recently proposed systems to overcome the problems caused by usage of neutral sections and can be applied in high-speed railway [12–16]. Shown in Figs. 1.4 and 1.5 are the typical circuit structures of co-phase traction power supply (Boost-Transformer, BT and Auto-Transformer, AT).

Three-phase ac power is transformed into two individual single-phase outputs through substation transformer. However, in contrast to conventional traction power supply, locomotive loads are connected across the same single-phase output. This can effectively eliminate the quantity of neutral sections due to less risk of phase mixing. Elimination of physical isolation helps to form a closed-loop power supply system with higher power reliability in case of section power failure. The isolation requirement and the quantity of isolators are less than conventional ones. Power

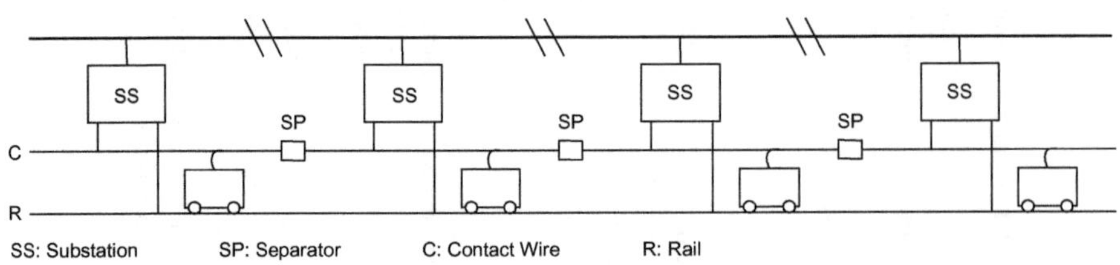

Fig. 1.4 Circuit diagram of co-phase traction power supply (BT)

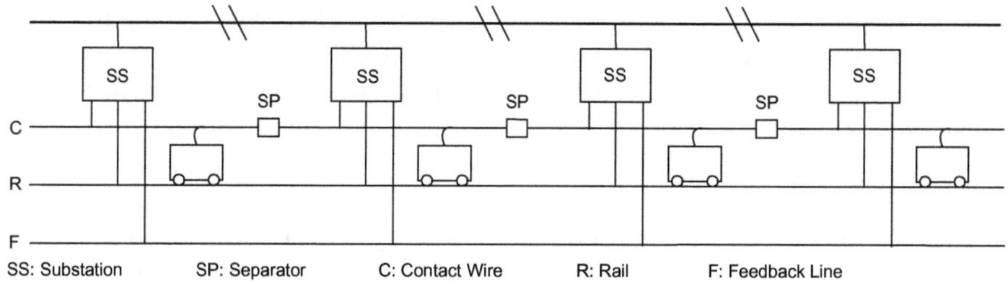

Fig. 1.5 Circuit diagram of co-phase traction power supply (AT)

loss due to neutral sections or isolators can therefore be effectively reduced. Based on the structure, co-phase traction power possesses numerous advantages over conventional ones.

1.4.2 Traction Power Quality Problems

It has been mentioned previously that traction load is dynamic, time-varying, and nonlinear that it may bring about various power quality problems simultaneously. Typical traction power supply power problems include (1) negative sequence and system unbalance; (2) power factor and reactive power; and (3) Harmonics and oscillating power. In fact, various standards have been issued on the power quality tolerance of different power systems. In this section, different existing problems of traction power supply are introduced together with their corresponding solutions. The IEEE and National standards are also quoted for reference.

System Unbalance and Negative Sequence

One major problem of traction power supply is negative sequence and system unbalance. Unbalance (voltage or current) occurs whenever negative sequence components are present in power systems. This may be due to connection of unbalanced loadings. In traction power supply, system unbalance is caused by unbalanced locomotive loading between two phases.

According to IEEE Standard 1159-2009 "IEEE Recommended Practice for Monitoring Electric Power Quality" [17], system unbalance is expressed as a ratio of the magnitude of negative sequence component to the magnitude of positive sequence component in percentage, as shown in (1.1).

$$\%\text{Voltage Unbalance} = \frac{|V_{\text{neg}}|}{|V_{\text{pos}}|} \times 100\% \tag{1.1}$$

There is hardly any written standard for system unbalance in IEEE. However, according to National Standard GB/T 15543-2008 "Quality of Electric Energy

Supply Admissible Three Phase Voltage Unbalance" [18], the maximum voltage unbalance allowed at point of common coupling (PCC) is 2% for long-term and 4% for short-term disturbances. This may also be the standard for traction power system unbalance tolerance.

Reactive Power and Power Factor

Another power quality concern in power system is the reactive power. According to the definition, considering a harmonic-free power system, the apparent power provided by the source may be divided into active and reactive power components [17]. Reactive power is drawn from the power source whenever the load is capacitive or inductive. With the exclusive usage of inductive loads such as induction motor, reactive power problem is a serious concern in power system. Shown in Fig. 1.6 is the famous power triangle showing the composition of active and reactive components in apparent power. The unit of apparent power S is VA, while the units for active P and reactive power Q are W and VAR, respectively. Under this condition, the relationship between P, Q, and S follows the expression shown in (1.2).

It should be noticed that real power contribution consumed by system load is the active power P only. Therefore, the amount of reactive power is preferred to be as little as possible. Defining the angle between active power P and apparent power S as θ, as reactive power Q decreases, the angle θ also decreases. With power factor defined as cos θ, system power factor is best to be near unity (1.0). When the power factor is close to unity, the amount of reactive power is near zero.

$$S^2 = P^2 + Q^2 \tag{1.2}$$

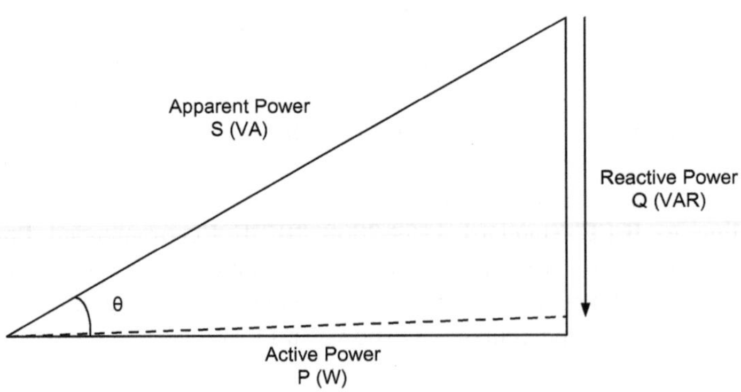

Fig. 1.6 Famous power triangle showing composition of active and reactive components in apparent power

Oscillating Power and Harmonics

Another serious problem of traction power supply is oscillating power and harmonics. Harmonics are generated whenever nonlinear load is connected. They are sometimes considered as the oscillating power between system source and load. With increasing usage of nonlinear devices in power system, harmonic problem is becoming more severe nowadays. This problem also exists in traction power supply since locomotives are nonlinear rectifier loads. The presence of harmonics in rotating machine can cause heat and copper loss which results in machine damage. In transformer, harmonics can bring about audible noise, copper, and stray flux losses. Therefore, harmonic compensation is required in traction power systems.

According to IEEE Standard 519-1992 "Recommended Practices and Requirements for Harmonic Control in Electrical Power Systems" [17], there are limits for harmonics tolerance in power systems at different voltage levels (Tables 1.2, 1.3 and 1.4). System performance should satisfy these standards.

Table 1.2 Basis for harmonic current limits in IEEE Std. 519-1992

Basis for Harmonic Current Limits		
SCR at PCC	Maximum individual frequency voltage harmonic (%)	Related assumption
10	2.5–3.0	Dedicated system
20	2.0–2.5	1–2 large customers
50	1.0–1.5	A few relatively large customers
100	0.5–1.0	5–20 medium size customers
1000	0.05–0.10	Many small customers

Table 1.3 Current distortion limits for generation distribution systems (120 V through 69,000 V) in IEEE Std. 519-1992

Current distortion limits for general distribution systems (120 V through 69,000 V)						
Maximum harmonic current distortion in percent of I_L						
Individual harmonic order (Odd harmonics)						
I_{sc}/I_L	<11	$11 \leq h < 17$	$17 \leq h < 23$	$23 \leq h < 35$	$35 \leq h$	TDD
<20[a]	4.0	2.0	1.5	0.6	0.3	5.0
20 < 50	7.0	3.5	2.5	1.0	0.5	8.0
50 < 100	10.0	4.5	4.0	1.5	0.7	12.0
100 < 1000	12.0	5.5	5.0	2.0	1.0	15.0
>1000	15.0	7.0	6.0	2.5	1.4	20.0

Even harmonics are limited to 25% of the odd harmonic limits above

Current distortions that result in a dc offset, e.g., half-wave converters, are not allowed

[a]All power generation equipment is limited to these values of current distortion, regardless of actual I_{sc}/I_L

where

I_{sc} = maximum short-circuit current at PCC

I_L = maximum demand load current (fundamental frequency component) at PCC

Table 1.4 Current distortion limits for generation distribution systems (69,001 V through 161,000 V) IEEE Std. 519-1992

Current distortion limits for general distribution systems (69,001 V through 161,000 V)						
Maximum harmonic current distortion in percent of I_L						
Individual harmonic order (Odd harmonics)						
I_{sc}/I_L	<11	$11 \le h < 17$	$17 \le h < 23$	$23 \le h < 35$	$35 \le h$	TDD
<20[a]	2.0	1.0	0.75	0.3	0.15	2.5
20 < 50	3.5	1.75	1.25	0.5	0.25	4.0
50 < 100	5.0	2.25	2.0	0.75	0.35	6.0
100 < 1000	6.0	2.75	2.5	1.0	0.5	7.5
>1000	7.5	3.5	3.0	1.25	0.7	10.0

Even harmonics are limited to 25% of the odd harmonic limits above
Current distortions that result in a dc offset, e.g., half-wave converters, are not allowed
[a]All power generation equipment is limited to these values of current distortion, regardless of actual I_{sc}/I_L
where
I_{sc} = maximum short-circuit current at PCC
I_L = maximum demand load current (fundamental frequency component) at PCC

Besides IEEE, National Standard has also set limits for harmonics in power system. The standard given in National Standard GB/T 14549-93 "Quality of Electric Energy Supply Harmonics in Public Supply Network" for harmonic tolerance is quoted in Fig. 1.7.

Rated Voltage (kV)	Rated Capacity (MVA)	Harmonic number and current tolerance (A)																							
		2	3	4	5	6	7	8	9	10	11	12	13	14	15	16	17	18	19	20	21	22	23	24	25
0.38	10	78	62	39	62	26	44	19	21	16	28	13	24	11	12	9.7	18	8.6	16	7.8	8.9	7.1	14	6.5	12
6	100	43	34	21	34	14	24	11	11	8.5	16	7.1	13	6.1	6.8	5.3	10	4.7	9.0	4.3	4.9	3.9	7.4	3.6	6.8
10	100	26	20	13	20	8.5	15	6.4	6.8	5.1	9.3	4.3	7.9	3.7	4.1	3.2	6.0	2.8	5.4	2.6	2.9	2.3	4.5	2.1	4.1
35	250	15	12	7.7	12	5.1	8.8	3.8	4.1	3.1	5.6	2.6	4.7	2.2	2.5	1.9	3.6	1.7	3.2	1.5	1.8	1.4	2.7	1.3	2.5
66	500	16	13	8.1	13	5.4	9.3	4.1	4.3	3.3	5.9	2.7	5.0	2.3	2.6	2.0	3.8	1.8	3.4	1.6	1.9	1.5	2.8	1.4	2.6
110	750	12	9.6	6.0	9.6	4.0	6.8	3.0	3.2	2.4	4.3	2.0	3.7	1.7	1.9	1.5	2.8	1.3	2.5	1.2	1.4	1.1	2.1	1.0	1.9

Fig. 1.7 Quotation of harmonic standard in National Standard GB/T 14549-93

1.4.3 *Existing Solutions for Traction Power Quality Problems*

It has been shown from practical statistics that traction power supply system power quality performance is usually far from standard and satisfactory. Various solutions have thus been proposed to overcome these problems. Different techniques used to solve the existing problems in traction power supply are listed below.

Solutions for System Unbalance and Negative Sequence

Various solutions have been developed to relieve system unbalance compensation in traction power supply. They are briefly described below.

Application of special three-phase to two-phase traction transformers

Special three-phase to two-phase traction transformers such as impedance matching balance, Scott and Wood-Bridge transformers, etc. can help in balancing three-phase source current at primary side. However, these transformers are expensive since they are specially made and are not preferred.

Power connection to different phases in rotating turns

Connecting the primary side of traction transformer to different source phases in rotating turns can also help to reduce the amount of negative sequences in the system. When the phase difference is 90° between two output terminals, or 120° between three output terminals, three-phase negative sequence components can be reduced. However, the system can never be completely balanced and system performance is still not satisfactory.

Installation of high-voltage and high-capacity power supply

Although high-voltage and high-capacity power supply tends to have higher ability to withstand unbalanced load, it usually involves high installation cost and is not preferred in practical application.

Installation of active compensation devices

Installation of active compensation devices such as Rail Power Compensator (RPC) can solve the unbalance problems. This method can provide unified power quality solution (refer to contents below) and has been widely used for solving traction power supply problems.

Solutions for Reactive Power and Power Factor

Besides unbalance, various solutions have been proposed for reactive power compensation or power factor correction.

Application of shunt capacitor

Shunt capacitor can be installed at substation traction side to relieve reactive power problems. However, its dynamic performance is poor and does not give satisfactory results for time-varying load.

Application of Active Compensation Devices

In contrast to shunt capacitor, compensators based on active power components can provide better dynamic and thus satisfactory performance. Various active compensators are therefore proposed for reactive power compensation (power factor correction) in traction power supply.

Solutions for Oscillating Power and Harmonics

As discussed previously, harmonic problem is also a great concern in traction power supply. Different solutions for harmonic compensation in traction power supply are summarized below.

Application of shunt LC filter

Shunt LC filter (series inductor and capacitor branch) can be installed at substation traction side to relieve harmonics problems. However, its dynamic performance is poor and does not give satisfactory results for time-varying load.

Application of multi-pulse rectifiers

Rectifying the ac input by multi-pulse rectifier can effectively reduce harmonics in the system. However, the number of components would increase and the rectifier size is larger.

Power capacity enhancement or change locomotive operation modes

The harmonics and reactive power problems could be relieved when the power capacity is increased. In addition, the same effect could be achieved by changing the locomotive operation modes. However, this usually involves higher installation cost and complicated operation procedures.

Application of shunt active compensation device

Shunt compensation device such as active power filter could be installed to compensate harmonics and reactive power. This method acts as a unified power quality solution and has been widely used.

A summary and comparison between various traction power quality problem solutions are shown in Table 1.5. As can be observed from the discussions of different traction power quality solutions above, various techniques have been proposed and used to solve corresponding problems. However, techniques suitable for solving one problem may not be suitable for solving others. On the other hand, active compensation devices can provide system unbalance, reactive power, and harmonic compensation simultaneously. Moreover, it has good dynamic performance and is therefore preferred over others. Thus, active power compensators have high potential to be a universal compensator for traction power supply.

Table 1.5 Comparisons and summary between various traction power quality problem solutions

	Unbalance solution	Reactive power solution	Harmonics solution	Unified solution	Remarks
Special transformer	✓	–	–	No	High cost
Rotating phase connection	✓	–	–	No	Never balanced
High voltage, capacity	✓	✓	–	No	High Cost
Shunt capacitor	–	✓	–	No	No dynamic compensation
Shunt LC filter	–	✓	✓	No	No dynamic compensation
Multi-pulse rectifier	–	–	✓	No	Large size
Active compensation devices	✓	✓	✓	Yes	Unified and comprehensive

1.5 Various Power Quality Compensators

Concerning the power quality problems mentioned above, researchers have proposed different compensators. Some of them have also been applied in traction power supply. They all have different characteristics and may aim at several specific power quality problems. Some of them have also been practically installed in power system. Initially, compensation used to be done using devices composed of passive components like fixed capacitor. However, due to fixed parameter setting, compensation is fixed and the dynamic compensation performance is poor. After the development of modern power electronics, active compensators based on Flexible Alternating Current Transmission System (FACTS) techniques have been widely studied [19]. Various active power compensators are then developed. Compared to passive compensators, active compensators can provide better dynamic compensation and performances.

Difference compensators developed are briefly discussed below. They are namely (1) fixed shunt capacitor bank [20]; (2) passive filter [21]; (3) Static Var Compensator (SVC) [22–24]; (4) Static Synchronous Compensator (STATCOM) [25–27]; (5) Dynamic Voltage Restorer (DVR) [28, 29]; (6) Unified Power Quality Compensator (UPQC) [30–32]; and (7) Hybrid Active Power Filter (HAPF) [33, 34]. Finally, comparisons are made among them.

1.5.1 Fixed Shunt Capacitor Bank

As introduced in the previous section, compensation is initially achieved using passive elements. Shunt capacitor bank is one of them. Loading such as traction locomotives are inductive, and capacitor bank may therefore be shunt connected at the point of common coupling (PCC) point for reactive power compensation, as shown in Fig. 1.8a. However, since the capacitor bank is fixed, the compensation capacity is also fixed such that dynamic compensation is not achievable. In addition, system unbalance and harmonic compensation cannot be provided by fixed capacitor bank.

1.5.2 Passive Filter

Another device that may provide harmonic compensation in addition to reactive power compensation is developed afterward. It is composed of series-connected LC (inductor and capacitor) branch and is commonly known as the passive filter, as shown in Fig. 1.8b. This name arises since inductors and capacitors are considered passive components. When the parameters are specifically selected, the impedance of LC branch at certain harmonic frequency is set to zero and can be considered as short circuit. This also provides a sink for harmonics such that the power source would be harmonic free. However, passive filter based on such design is usually effective for one certain harmonic only. For example, in common 6-diode rectifier load, the harmonics are usually concentrated around 5th and 7th harmonics, and

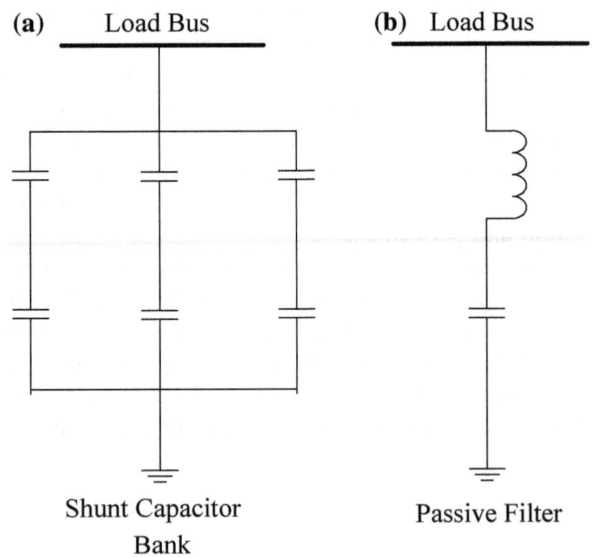

Fig. 1.8 Two passive power quality compensators: **a** Shunt capacitor bank; **b** Passive filter

reduction of a single harmonic content is not an effective compensation technique. In addition, system with passive filter may suffer from resonance problem which may cause harmonic problem more serious than before.

1.5.3 Static Var Compensator (SVC)

After the development of modern power electronics, electronic switching technique emerges and brings changes to power quality compensation. The FACTS device family is then widely researched. Static Var Compensator (SVC) is one member of the shunt FACTS device family members that are used for reactive power compensation and voltage regulation. Some typical circuit structures of SVC are shown in Fig. 1.9. SVC is developed by replacing the mechanical switches in traditional shunt reactor and capacitor by thyristors. Typical SVCs include Thyristor-Controlled Reactor (TCR), Thyristor-Switched Capacitor (TSC) or combined SVC such as Fixed Capacitor–Thyristor-Switched Reactor (FC-TSR). SVC can act as reactive power source or sink according to the load conditions. For example, when the load is capacitive (leading), SVC is switched to reactor mode and absorbs reactive power from the system, thus lowering system voltage; on the other hand, when the load is inductive (lagging), SVC is switched to capacitor

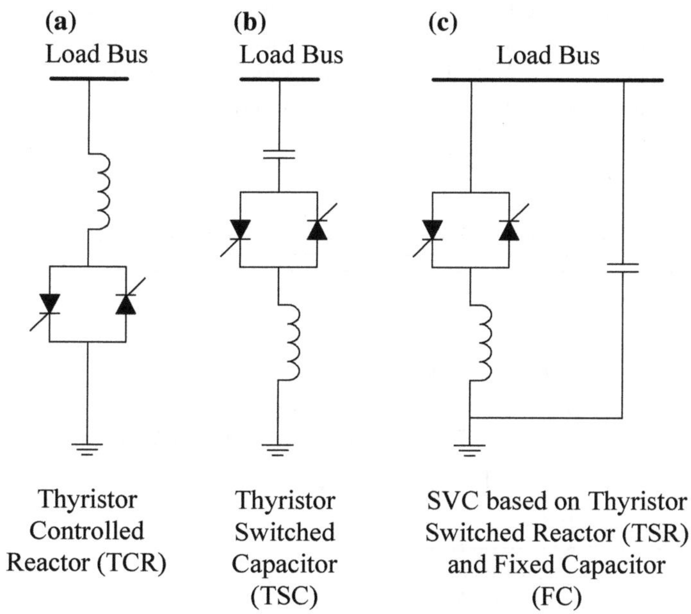

Fig. 1.9 Typical circuit schematics of Static Var Compensator (SVC): **a** TCR; **b** TSC; and **c** FC-TSR

mode and generates reactive power to the system, thus maintaining system voltage. However, SVC cannot provide fast dynamic compensation performance and is not preferred when the load is dynamic and time-varying. In addition, the compensation power of SVC is relatively less and limited compared to other advanced FACTS devices.

1.5.4 Static Synchronous Compensator (STATCOM)

Although SVC provides better compensation than conventional compensators, its dynamic performance is still far from satisfactory under dynamic load. Controllable electronic switches such as IGBT then contribute largely to the development of advanced compensators such as Static Synchronous Compensator (STATCOM) or Active Power Filter (APF). STATCOM or active power filter is sometimes known as an advanced version of SVC. Typical STATCOM/APF circuit schematic is shown in Fig. 1.10.

It is connected between power source and nonlinear load for compensation. STATCOM and APF share the same circuit structure. When the compensation device is used to compensate harmonics only, it is regarded as APF; when reactive compensation is also considered, it is regarded as STATCOM. The term "static" refers to non-rotational device that is different from traditional compensation machine that may generate audible noise. STATCOM/APF is composed of a dc link and a voltage source inverter (VSI). The VSI is used to convert dc link power into ac so as to compensate harmonics, active or reactive power as desired. Load harmonics are generated by APF during compensation. When the load is capacitive (leading), the inverter voltage is less than the system voltage and a lagging current

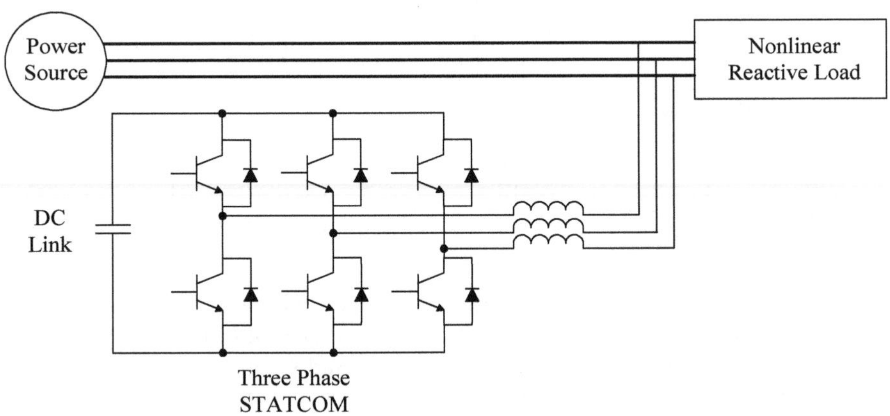

Fig. 1.10 Typical circuit schematic of static synchronous compensator (STATCOM) or Active Power Filter (APF)

is injected (absorbing reactive power); on the other hand, when the load is inductive (lagging), the inverter voltage is higher than system voltage and a leading current is injected (generating reactive power). The dynamic compensating power of STATCOM/APF is higher than SVC and is mainly used to compensate current harmonics and reactive power.

1.5.5 Dynamic Voltage Restorer (DVR)

Aiming at different compensation targets, Dynamic Voltage Restorer (DVR) is a series FACTS compensation device that is used to compensate voltage variations such as voltage sag. Typical DVR circuit schematic is shown in Fig. 1.11.

The structure of DVR is similar to that of STATCOM, only that the inverter is connected in series with the system through transformers to compensate load voltage variations. Compared to other compensators, DVR does not provide any direct harmonic or reactive power compensation. Instead, voltage variation compensation is its main concern with regard to power quality. It is introduced here since some advanced version has been developed for universal compensator afterward.

1.5.6 Unified Power Quality Compensator (UPQC)

As some power system load may be critical and requires high power quality and voltage stability, the compensators mentioned above may not be appropriate as a universal compensator by oneself. Researchers have then explored other new

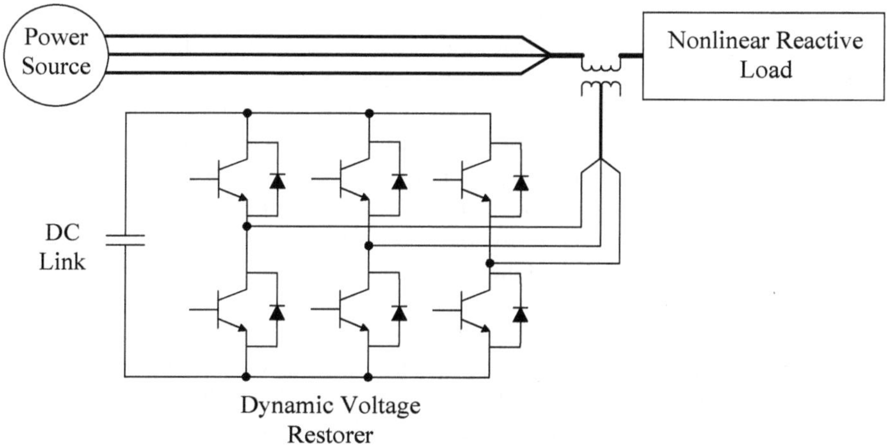

Fig. 1.11 Typical circuit schematic of Dynamic Voltage Restorer (DVR)

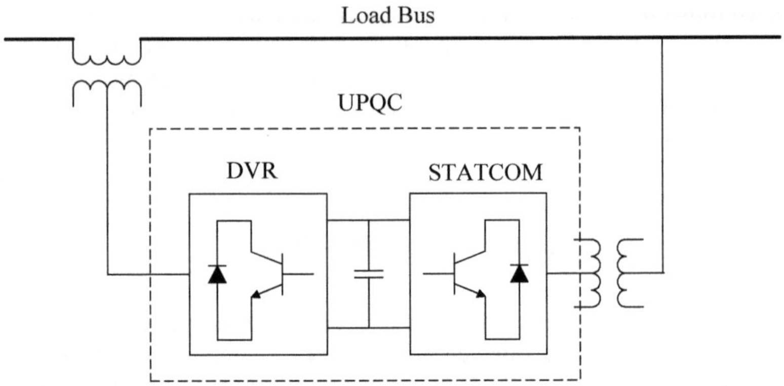

Fig. 1.12 Typical circuit schematic of Unified Power Quality Controller (UPQC)

universal compensation device. Unified Power Quality Controller (UPQC) is one of them. It is a combination of DVR and STATCOM connected through the same dc link. Typical UPQC structure is shown in Fig. 1.12. As introduced above, DVR is suitable for compensating voltage variations, while STATCOM is suitable for compensating current harmonics. UPQC is considered as a universal solution for critical load compensation where high power quality is required. However, as can be observed from its structure, the number of switching devices is doubled compared to previous compensators. UPQC is therefore not suitable for practical application when system load power quality requirement is not that difficult.

1.5.7 Hybrid Active Power Filter (HAPF)

Although active compensators such as STATCOM/APF can provide better dynamic performance than passive compensators like passive filter, the cost of active compensator installation is still higher than passive ones. Researchers have therefore explored another compensator, Hybrid Active Power Filter (HPAF), which combines both advantages of active and passive compensators. HAPF refers to compensation devices that are composed of both active and passive compensators. HAPF is proposed to reduce the workload (and thus device rating) of VSI in STATCOM and DVR. Different hybrid APF structures have been proposed and three typical circuit schematics of them are shown in Fig. 1.13. Different targets can be achieved using different topologies. For example, the compensation voltage rating of APF can be greatly reduced using shunt hybrid APF. It is definitely true that HAPF can provide dynamic compensation with lower cost under certain load condition. However, it is designed mainly for inductive load. With increasing usage of capacitive load such as energy saving devices, HAPF may not be a good compensator under all conditions.

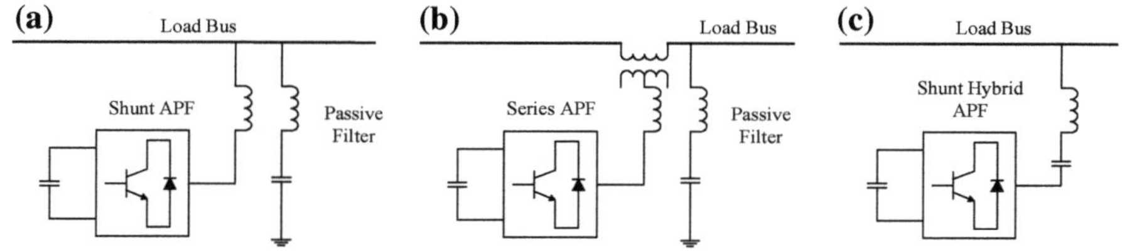

Fig. 1.13 Typical structures of Hybrid Active Power Filter (Hybrid APF)

1.5.8 Compensators Applied in Traction Power

Some power quality compensators have also been applied in traction power supply. Some of them are discussed in details below.

Shunt Capacitor Bank

In traction power supply, shunt capacitor is installed every certain distance along the traction power supply line to provide reactive power compensation. However, as mentioned, reactive power compensation of shunt capacitor bank is fixed and cannot provide dynamic compensation. It is not preferred for traction power quality compensation.

Passive Filter

Passive filters have also been applied to traction power supplies to filter certain order harmonics and to provide reactive power compensation. However, as discussed, passive filter is composed of fixed inductor and capacitor, the reactive compensation power that can be provided is also fixed, and may not provide satisfactory compensation performance during load variations. Furthermore, resonance may occur such that the harmonic distortions are even larger.

Static Var Compensator (SVC)

Besides passive compensators, compensators based on active switching components have also been used in providing traction power quality compensation. SVC is one example and can provide better dynamic compensation performance. However, the compensation range is still limited and SVC may inject harmonics into traction power grid and make compensation performance less satisfactory.

STATCOM

With the development of power electronics, FACTS compensation devices such as STATCOM have been used to provide more unified and comprehensive power quality compensation in traction power. STATCOM is one of them. The STATCOM was installed at the three-phase primary source grid of traction transformer. Power quality like reactive power and harmonic compensation is thus provided directly to the source grid. Harmonics are therefore still present in transformer and may damage the device (Fig. 1.14).

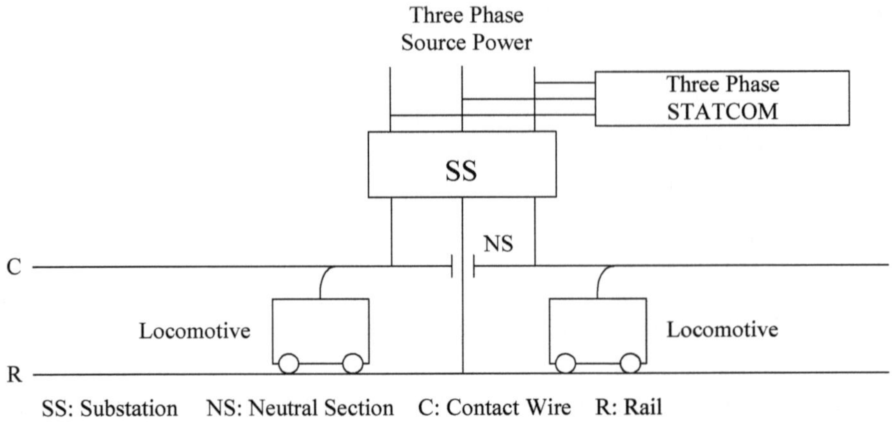

Fig. 1.14 Illustration diagram to show the installation of three-phase STATCOM to provide power quality compensation in traction power supplies

RPC

The railway power quality conditioner (RPC) was developed by a Japan scholar and had been used in railway traction power supply. Different from STATCOM, the RPC is installed across transformer output and power quality compensation is provided from the secondary side to the primary side, as shown in Fig. 1.15. This power quality compensation device is more preferred since harmonics are compensated at secondary side and will not pass through the traction transformer. However, the control derivation is complicated and is different from STATCOM.

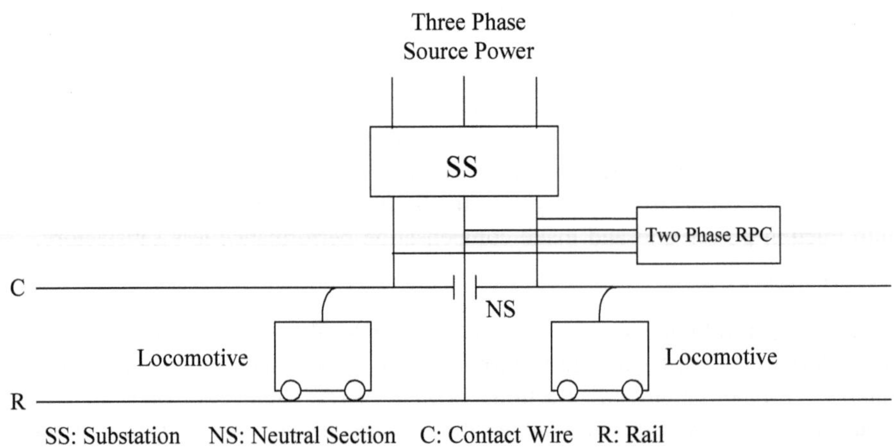

Fig. 1.15 Illustration diagram to show the installation of two-phase RPC to provide power quality compensation in traction power supplies

1.5.9 Comparisons Among Various Compensators

It is always desirable to select the best power compensator. However, it is difficult to draw a conclusion since the compensators all aim at different compensation targets. A universal compensator usually possesses various drawbacks such as high installation cost. Shown in Tables 1.6 and 1.7 are the comparisons between different FACTS compensation devices (for the same compensation capacity) introduced above.

Table 1.6 Comparisons among various mentioned power quality compensators (1)

	Shunt capacitor	Passive filter	SVC	STATCOM/APF
Compensation target	Voltage regulation, reactive power	Current harmonics	Voltage regulation	Voltage regulation, current harmonics
Response time	NA	NA	Slow	Fast
Dynamic response	Poor	Poor	Poor	Better
Compensating power	Low and fixed	Lower	Lower	Higher
Device rating	Higher	Lower	Higher	Higher
Size	Large	Smaller	Largest	Smaller
Control	NA	NA	Fair	Easy
Price	Lower	Lower	Lower	Higher

Table 1.7 Comparisons among various mentioned power quality compensators (2)

	DVR	UPQC	Hybrid APF
Compensation target	Voltage sag compensation	Current harmonics, voltage sag compensation	Current harmonics (reduced APF rating)
Response time	Fast	Fast	Fast
Dynamic response	Better	Better	Better
Compensating power	Higher	Higher	Higher
Device rating	Higher	Higher	Lower
Size	Smaller	Larger	Medium
Control	Easy	Complicated	Fair
Price	Higher	Highest	Higher

1.6 Recent Research Developments on Traction Power Supply System and Its FACTS Compensation Devices

Next, recent researches on traction power supply system and its FACTS compensation devices are briefly introduced.

1.6.1 Recent Research on Traction Power Supply System

Many recent researches on traction power supply system are focused on new co-phase traction power supply system, which has been introduced previously for advantages of elimination of neutral sections and higher transformer utilization ratio, etc. The co-phase traction power supply system is first proposed by a research group at Southwest Jiaotong University in China. The circuit schematics of a typical co-phase traction power supply system being investigated is shown in Fig. 1.16. Relevant results and theories can be found in [12, 13]. As introduced, the world's first co-phase traction device has already been put in trial operation in KunMing MeiShan substation. Important data and analysis have also been presented in [35]. However, the power quality compensator used in the system is

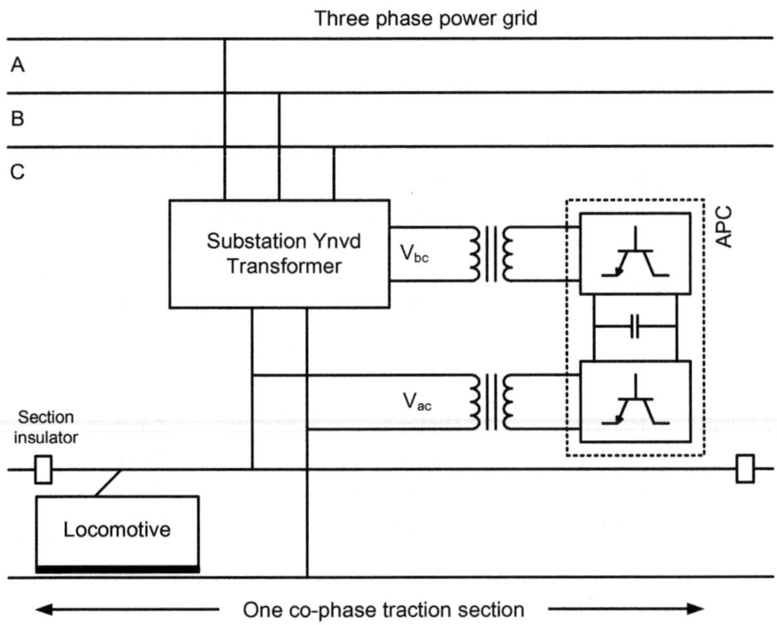

Fig. 1.16 Typical structure of co-phase traction power supply system proposed by Southwest Jiaotong University

inductively coupled and causes high operation voltage and device rating requirement. This induces high cost and limits co-phase traction power development.

Besides proposing co-phase traction power supply, the research group has also proposed a co-phase traction power supply system based on three-phase to single-phase power electronics converter, as shown in Fig. 1.17. According to their study idea, the converters can act as a solid-state transformer and can work independently but interact with others just like grid-connected solid-state transformer. A new control algorithm for current and power sharing is also developed. It is claimed that the frequency and voltage droop control can improve the system stability of multi-inverter connected grid. However, in this proposed structure, the power system supply reliability is highly dependent on the converter. Once the converter fails, the power supplied to locomotive load will be unstable. Furthermore, step-up and step-down transformers are required that the number of components is not less compared to traditional co-phase system shown in Fig. 1.16. Therefore, this structure may also not be beneficial for co-phase traction power development.

The emergence of co-phase traction power supply system can cause advances in high-speed railway. If its initial cost can be reduced, it helps to further enhance its development.

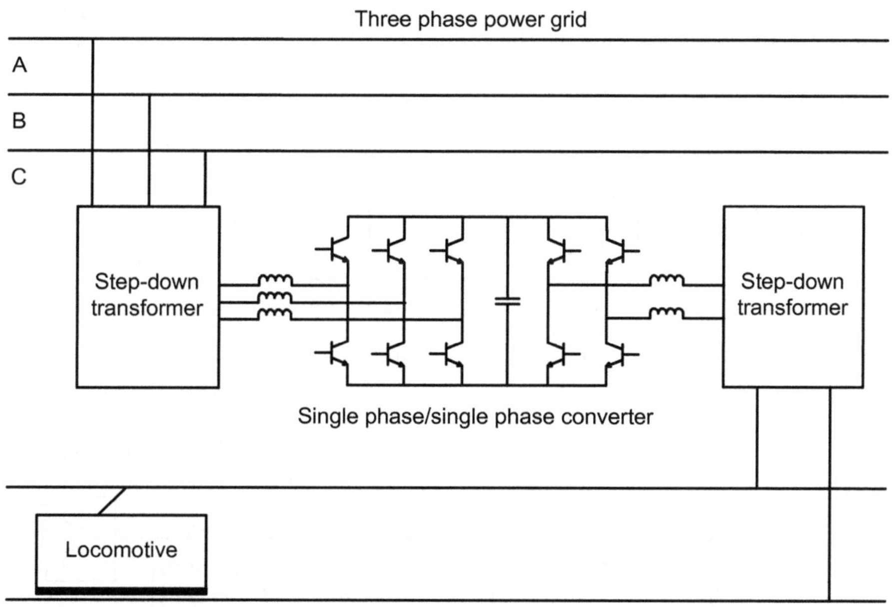

Fig. 1.17 Circuit schematics of enhanced co-phase traction based on three-phase to single-phase power electronics converter

1.6.2 *Recent Researches on Traction FACTS Compensation Devices*

As discussed, compared to traditional passive compensation devices, FACTS compensation devices based on active components provide better dynamic compensation performance and are preferred. Researchers have also done many researches concerning the usage of FACTS compensation devices in traction power supply. Traditionally, SVC is used for traction power quality compensation. For instance, in [36], the compensation performance using TSR SVC for voltage regulation of a 25 kV traction is explored (Fig. 1.18). However, its dynamic performance is poor and compensation results are far from satisfactory when the load is varying. Furthermore, the size of high power SVC is large and occupy huge area. Although active power filter (APF) is then proposed to provide fast and dynamic response, the device rating of APF is still too high and includes high cost. Therefore, in [37], a traction power compensation device based on hybrid structure is proposed for lower device rating and initial cost (Fig. 1.19). Besides APF, there are also different proposed topologies of STATCOM in traction power compensation. The commonly used two-phase STATCOM is known as Railway Power Compensator (RPC) and is proposed by scholars in Japan [38] (Fig. 1.20). It is used to provide power quality compensation at source side by providing compensation power from the secondary side. The two-phase STATCOM is also adopted for power quality compensation by the co-phase traction power research group in

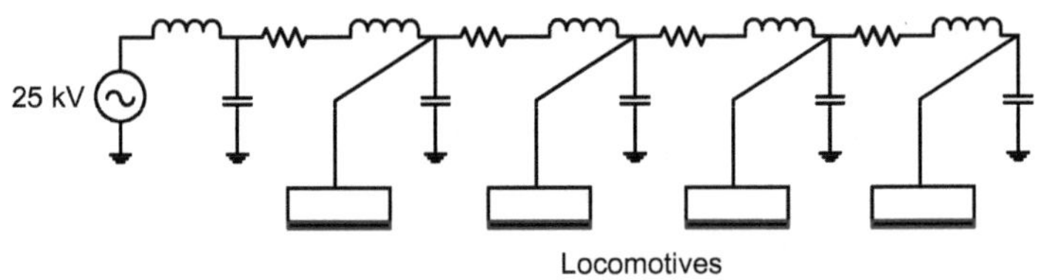

Fig. 1.18 Application of TSR SVC for voltage regulation of traction power

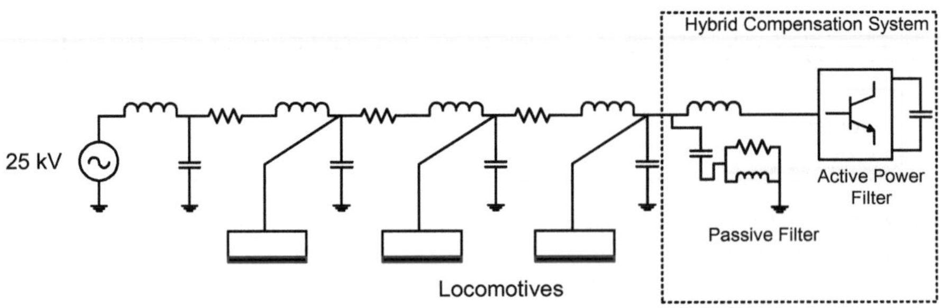

Fig. 1.19 Application of hybrid active power filter in traction power supply

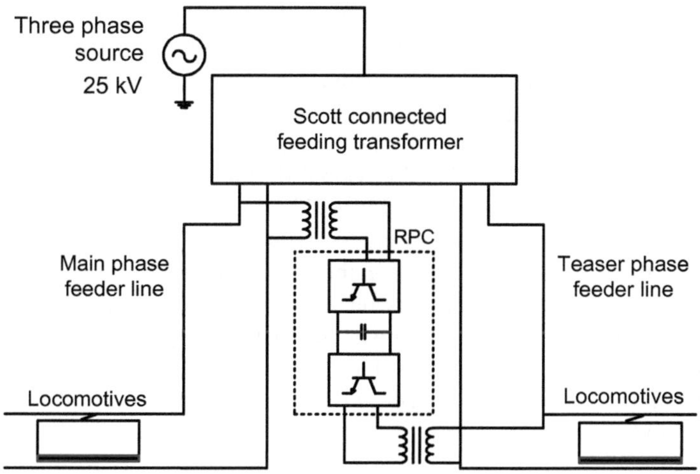

Fig. 1.20 Application of two-phase STATCOM in traction power is known as RPC in Japan

China. On the other hand, three-phase STATCOM for compensation at substation secondary side is also proposed for fewer components and better application of three-phase instantaneous pq theory (Fig. 1.21). However, active and reactive flow in the proposed structure becomes complicated and specially made transformer is required. This results in drawbacks such as increase in control complexity and cost.

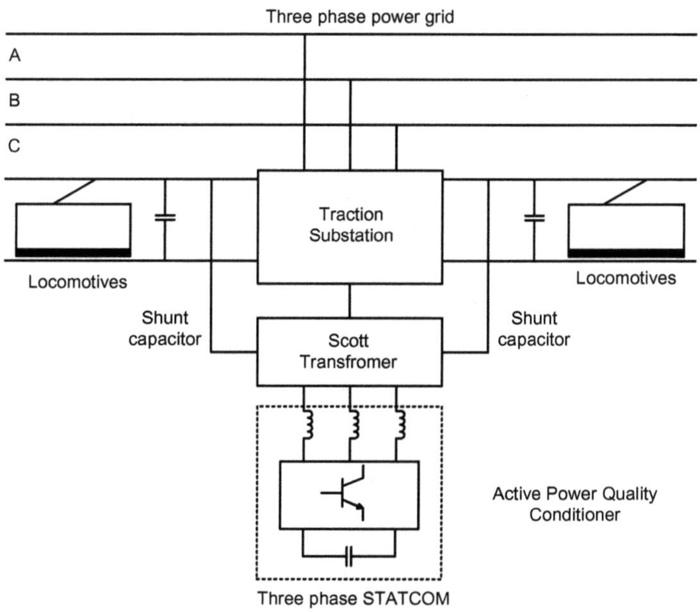

Fig. 1.21 Three-phase STATCOM proposed for traction compensation

In high power applications, compensation device with multilevel structure is preferred. Different structures of multilevel topology (series and parallel) have been proposed. For example, in [39], an individual dc link cascaded multilevel structured STATCOM is proposed (Fig. 1.22). However, two transformers are required. The cost of transformers in high power application is usually high and better be avoided. Based on this reason, in [40, 41], individual dc link cascaded chained multilevel structured STATCOM is proposed (Fig. 1.23). In this structure, one transformer can be omitted and installation cost can be reduced.

Recently, there are also researches on the analysis of application using traditional techniques. For example, researchers have analyzed the effect of different installation positions of SVC in traction power supplies in [42]. Although the researches can help to analyze the location which SVC can be installed to have the best performance, they have less impact on further enhancing the development of high-speed railway traction power.

1.6.3 Research Development of Co-phase Traction Power with Railway HPQC

Since 2008, the research team at University of Macau is the first one in the world which has proposed and performed in-depth investigations on co-phase traction power with hybrid capacitive-coupled Railway HPQC for application in high-speed railway power supply, as shown in Fig. 1.24. It has the advantages of co-phase traction power, but can solve the problem of high operation voltage in conventional

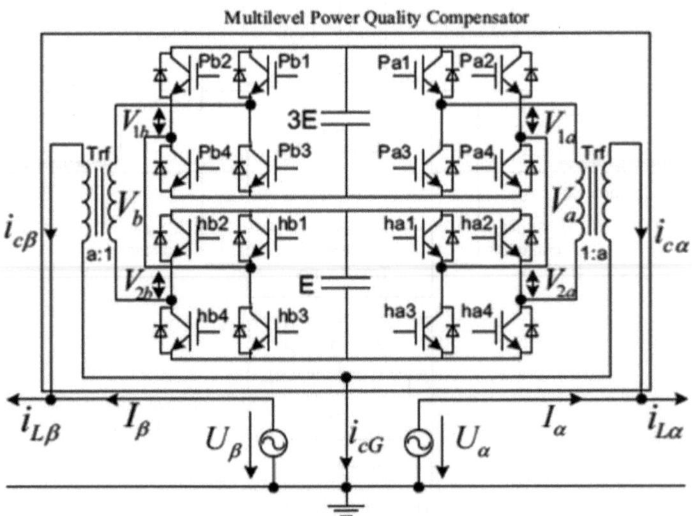

Fig. 1.22 Application of multilevel STATCOM in traction power compensation

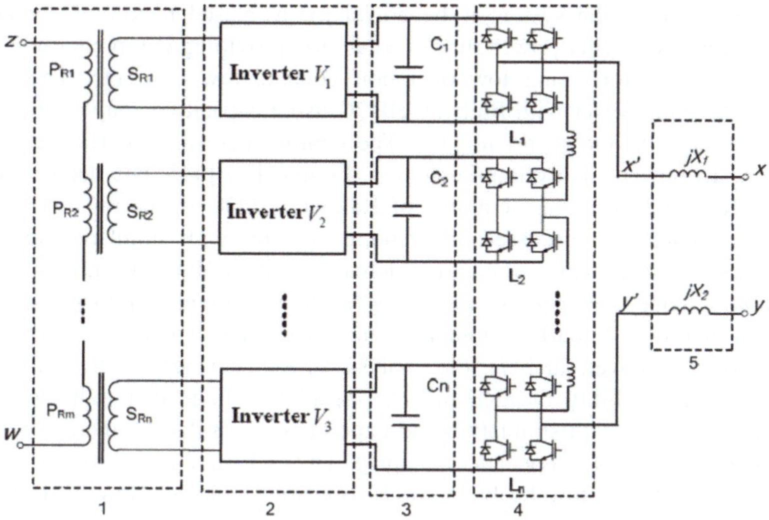

Fig. 1.23 Proposed individual dc link cascaded multilevel structure STATCOM

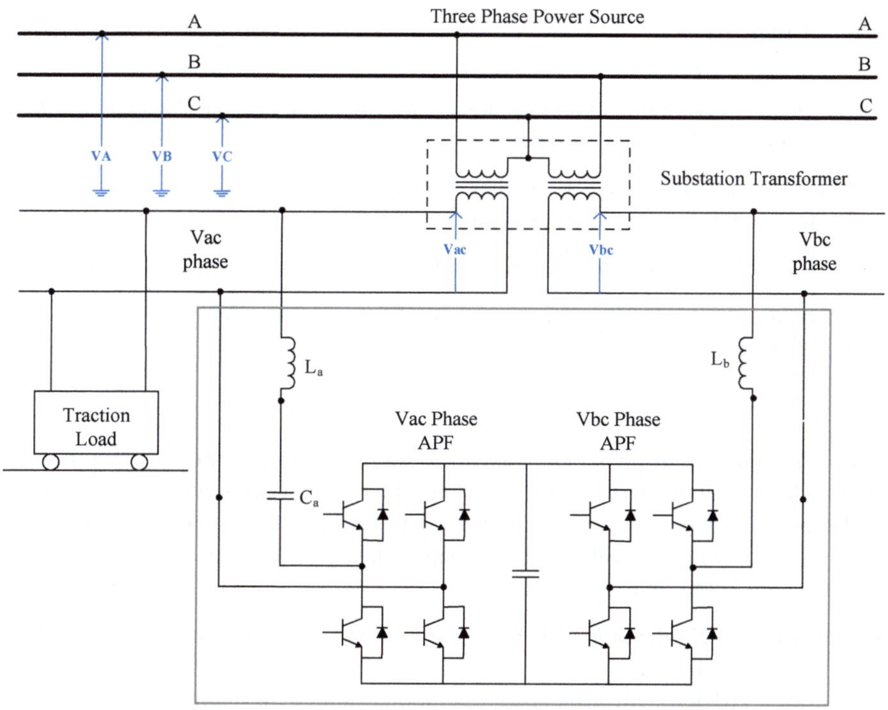

Fig. 1.24 Circuit structure of co-phase traction power with Railway HPQC

system with RPC. The system is beneficial for increased maximum locomotive speed achievable, higher power supply reliability, providing comprehensive power quality compensation using low operation voltage, low cost, etc. The idea of co-phase traction power with Railway HPQC using capacitive-coupled impedance is first revealed to the world in [43]. The control algorithm of Railway HPQC according to power quality requirement is discussed in [44]. The detailed system design procedure and considerations of Railway HPQC operation for minimum operation voltage are discussed and reported in [45] with simulation and result verification. More consideration and systematic analysis about further reduction of operation voltage under harmonic compensation through parameter design is explored in [46]. Similarly, simulation and experimental results are shown for verification. Moreover, in [47], the analysis of enhancement in Railway HPQC compensation capability through raising operation voltage to a suitable level is presented, together with simulation and experimental results. Furthermore, in [48, 49], the application of "impedance-mapping" and "adaptive dc control" techniques is used to maintain a low Railway HPQC operation voltage. New idea of partial compensation for possible reduction in operation voltage and cost in Railway HPQC is also reported and shared in [50, 51].

1.7 Summary

The co-phase traction power supply has the following advantages, which is suitable for high-speed railway development:

- Elimination of neutral sections that limit locomotive velocity;
- Higher transformer utilization ratio.

Next, the Railway HPQC with hybrid LC-coupled structure is advantageous for

- Providing unified solution for power quality compensation;
- Lower inverter voltage ratings;
- Lower device ratings and cost;
- Lower operation loss.

In this book, the co-phase traction power supply system is modeled first. In order to provide complete analysis of the system, the compensation algorithm according to the power quality criteria is derived. The design of co-phase traction power supply system with the following different Railway HPQC based on various considerations is being analyzed:

- Railway HPQC developed based on minimum operation voltage under fixed designed rated load;
- Railway HPQC with enhanced operation voltage to increase compensation capability during load variations;

- Railway HPQC with impedance-mapping technique to provide satisfactory compensation within load variation range with lower operation voltage and coupled capacitance;
- Railway HPQC with adaptive dc link voltage control algorithm to automatically adjust operation voltage to adjust the compensation capability for load variations.

PSCAD simulations of the system are performed based on the practical locomotive power data obtained from China WuQing substation. A laboratory-scaled hardware system and prototype is also constructed to verify the performance of the system. Comparisons are also made between system with Railway HPQC and conventional one with RPC to show the advantages of providing similar system performance with lower operation voltage. Finally, further possible investigations and research directions are discussed.

1.8 Book Organization

The electric traction power supply system for unified power applications based on Railway HPQC is introduced in this book. In this chapter, the background about traction power and co-phase traction power supply structure is given. In order to understand more about the topic, the development of electrified railway power is reviewed. The conventional co-phase traction power supply structure with RPC and new one with Railway HPQC are then introduced. It is shown that co-phase traction power possesses numerous advantages over conventional ones. Besides, different power quality problems are traction power that are discussed together with existing solutions. Various compensators are also reviewed so as to understand their characteristics. Finally, recent research developments on traction power compensation are reviewed.

In Chap. 2, the control algorithm for Railway HPQC is being derived and developed. Three major power quality problems in traction power, system unbalance, reactive power, and harmonics are being modeled first. Then, the required compensating power of the Railway HPQC is derived according to the power quality requirement. The entire control algorithm is then presented based on the analysis and the instantaneous pq theory [52]. Further comparison of the system with Railway HPQC is compared with conventional one with RPC to show the advantages.

In Chap. 3, the whole system is being analyzed and the Railway HPQC design procedure based on minimum operation voltage requirement at fixed rated load is developed. For simplicity, the system is first divided into linear combination of fundamental and harmonic models. Analyses are performed on each model so as to obtain an appropriate parameter design. In order to reduce the installation cost, the parameter is designed for minimum HPQC operation voltage rating. Simulations and experimental results are also provided to verify the system performance under such parameter design.

In Chap. 4, since locomotive loads are mostly dynamic [53], further investigations are done with the relationship between Railway HPQC operation voltage and loading conditions. The analysis shows that full compensation can be provided for a range of load conditions, known as the compensation range. The mathematical relationship between HPQC operation voltage rating and compensation range is explored, together with the load condition limit for full compensation. Three different methods of Railway HPQC design for providing satisfactory compensation during load variations are developed. They are namely (a) Railway HPQC with enhanced operation voltage to increase the compensation capability, (b) Railway HPQC based on impedance-mapping technique to provide satisfactory compensation performance within load variation range, and (c) Railway HPQC with adaptive dc link voltage control algorithm to adjust its compensation capability during load variations. The relationship between Railway HPQC parameters coupled impedance, operation voltage, and compensation capability is first investigated. The comprehensive design procedures of the three methods are then presented after analysis. Each of them has different advantages, and the criteria for selection are also discussed. Similarly, simulation and experimental results are provided to support the developed theories.

In Chap. 5, partial compensation control algorithm for Railway HPQC is discussed. Further reduction in Railway HPQC can be archived by modification of the parameters in the control. The effect of modified control on Railway HPQC performance is investigated. Simulation and experimental results are also presented to show its effectiveness.

In Chap. 6, details of hardware construction and experimental results are presented. Hardware prototype is constructed in order to verify the system performance of the co-phase traction power supply with Railway HPQC of different designs.

Finally, in Chap. 7, a conclusion of the whole project is provided. It acts as a summary of the whole book. Possible further investigations and research direction are also introduced and discussed.

References

1. "Revising the Long and Mid-Term Plan of the China's railway", adjusted at 2008.
2. "2009 Construction Plan for Phase 1 of LRT System", Macau 2009, http://www.git.gov.mo/lrt1_2009/en_index.html.
3. "Research Progress for the Macao LRT", Macau 2010, http://www.git.gov.mo/en/history.aspx.
4. Fan Chun-lei; Wu Guang-ning; Zhang Xue-yuan; Wang Yun-fei, "The measurement and simulation research of electrified railway harmonic", International Conference on Condition Monitoring and Diagnosis, 2008, pp: 769–772.
5. Capasso, A.; Lamedica, R.; Sangiovanni, S.; Maranzano, G.; and Prudenzi, A., "Harmonics and PQ events monitoring in an electrified metro-transit system", 10th International Conference on Harmonics and Quality of Power, 2002, vol. 2, pp: 441–446.
6. Ribeiro, P.F., "An overview of power quality problems in transportation and isolated power systems", IEEE Power Engineering Society Summer Meeting, 2001, vol. 1, pp: 196–197.

7. Mellitt, B.; Allan, J.; Shao, Z.Y.; Johnson, W.B.; Hooper, A.; and Denley, M.R., "Harmonic characteristics of traction loads on New Zealand's newly electrified North Island line", 10th International Conference on Electricity Distribution, 1989, vol. 5, pp: 392–396.
8. Griffin, A.J., "Methods of improving the voltage regulation on 25 kV electric railways", International Conference on Main Line Railway Electrification, 1989. 1989, pp: 252–259.
9. Burdett, S.; Allan, J.; Mellitt, B.; Taufiq, J.A., "A study of power factor correction techniques for high power AC locomotives", International Conference on Main Line Railway Electrification, 1989. 1989, pp: 363–367.
10. Gotham, D.J.; and Heydt, G.T., "Power flow control and power flow studies for systems with FACTS devices", IEEE Transactions on Power Systems, 1998, vol. 13, issue: 1, pp: 60–65.
11. Steimel, A., "Electric railway traction in Europe", IEEE Industry Applications Magazine, 1996, vol. 2, issue: 6, pp: 6–17.
12. Z. Shu, L. Xie and Q. Li, "Single-phase back-to-back converter for active power balancing, reactive power compensation and harmonic filtering in traction power system", IEEE Trans. Power Electronics, Issue 99, 2010.
13. Zeliang Shu; Shaofeng Xie; and Qun-zhan Li, "Development and Implementation of a Prototype for Co-phase Traction Power Supply System", *Asia-Pacific Power and Energy Engineering Conference (APPEEC),* 2010, pp: 1–4.
14. Minwu Chen; Qunzhan Li; and Guang Wei, "Optimized Design and Performance Evaluation of New Cophase Traction Power Supply System", *Asia-Pacific Power and Energy Engineering Conference (APPEEC),* 2009, pp: 1–6.
15. Zhou Fulin, Li Qunzhan, Qiu Daqiang, "Co-phased traction power system based on balanced transformer and hybrid compensation", in Power and Energy Engineering Conference, 2009, APPEEC 2009, Asia-Pacific, 2009, pp. 1–4.
16. Zeng Guohong; Hao Rongtai, "Analysis and design of an active power filter for three-phase balanced electrified railway power supply system", *The Fifth International Conference on Power Electronics and Drive Systems, PEDS 2003,* 2003, vol. 2, pp: 1510–1513.
17. IEEE Standard 519-1992, "IEEE Recommended Practices and Requirements for Harmonic Control in Electrical Power Systems".
18. National Standard GB/T 15543-2008, "Quality of Electric Energy Supply Admissible Three Phase Voltage Unbalance".
19. Gotham, D.J.; and Heydt, G.T., "Power flow control and power flow studies for systems with FACTS devices", IEEE Transactions on Power Systems, 1998, vol. 13, issue: 1, pp: 60–65.
20. Andrei, R.G.; Kaushik, R.R.; Reinaker, R.W., "Bridge capacitor bank design and operation", *IEEE Transactions on Power Delivery*, vol. 11, issue: 1, 1996, pp: 227: 233.
21. Peng, F.Z.; Akagi, H.; Nabae, A, "A new approach to harmonic compensation in power systems", *Industry Applications Society Annual Meeting, 1988.*, vol. 1, 1988, pp: 874–880.
22. Gelen, A.; Yalcinoz, T., "The behavior of TSR-based SVC and TCR-based SVC installed in an infinite system", *IEEE 25th Convention of Electrical and Electronics Engineers in Israel,* 2008, pp: 120–124.
23. Li Zhang; Jie Lou; Qingmin Li; Qingquan Li, "Analysis of the Conducted Interfering Mechanism within SVC based substations", *International Symposium on Electromagnetic Compatibility,* 2007, pp: 440–444.
24. Gelen, A.; Yalcinoz, T., "Analysis of TSR-based SVC for a Three-Phase System with Static and Dynamic Loads", *International Conference on Electrical Engineering,* 2007, pp: 1–6.
25. Jianye Chen; Shan Song; Zanji Wang, "Analysis and Implement of Thyristor-based STATCOM", *International Conference on Power System Technology, PowerCon2006,* pp: 1–5.
26. de Assi, T.M.L.; Watanabe, E.H.; Pilotto, L.A.S.; and Sollero, R.B., "A new technique to control reactive power oscillations using STATCOM", *10th International Conference on Harmonics and Quality of Power,* 2002, vol. 2, pp: 607–613.
27. Sen, K.K.; "STATCOM-STATic synchronous COMpensator: theory, modeling and applications", *IEEE Power Engineering Society 1999 Winter Meeting,* 1999, vol. 2, pp: 1177–1183.

28. Huann-Keng Chiang; Bor-Ren Lin; and Kuan-Wei Wu, "Study of dynamic voltage restorer under the abnormal voltage conditions", *International Conference on Power Electronics and Drives Systems, PEDS 2005,* 2005, vol. 1, pp: 308–312.
29. Nielsen, J.G.; Newman, M.; Nielsen, H.; and Blaabjerg, F., "Control and testing of a dynamic voltage restorer (DVR) at medium voltage level", *IEEE Transactions on Power Electronics,* 2004, vol. 19, issue: 3, pp: 806–813.
30. Khadkikar, V.; Chandra, A.; Barry, A.O.; and Nguyen, T.D., "Conceptual Study of Unified Power Quality Conditioner (UPQC)", *IEEE International Symposium on Industrial Electronics,* 2006, vol. 2, pp: 1088–1093.
31. Ghosh, A.; Jindal, A.K.; and Joshi, A., "A unified power quality conditioner for voltage regulation of critical load bus", *IEEE Power Engineering Society General Meeting,* 2004, vol. 1, pp: 471–476.
32. Fujita, H.; and Akagi, H., "The unified power quality conditioner: the integration of series and shunt active filters", *IEEE Transactions on Power Electronics,* 1998, vol. 13, issue: 2, pp: 315–322.
33. Wu Jian; He Na; and Xu Dianguo, "A 10 kV shunt hybrid active filter for a power distribution system", *Twenty-Third Annual IEEE Applied Power Electronics Conference and Exposition, APEC2008,* 2008, pp: 927–932.
34. Jain, Taruna; Jain, Shailendra; and Agnihotri, Ganga, "Comparison of topologies of hybrid active power filter", *IET-UK International Conference on Information and Communication Technology in Electrical Sciences (ICTES2007),* 2007, pp: 503–509.
35. Zeliang Shu; Shaofeng Xie; Ke Lu; Yuanzhe Zhao; Xiaoqiang Nan; Daqiang Qiu; Fulin Zhou; Sibin Gao; Qunzhan Li, "Digital Detection, Control, and Distribution System for Co-Phase Traction Power Supply Application", *IEEE Transactions on Industrial Electronics,* 2013, vol: 60, issue: 5, pp: 1831–1839.
36. Inzunza, R.; and Akagi, H., "A 6.6-kV transformer less shunt hybrid active filter for installation on a power distribution system", *IEEE 35th Annual Power Electronics Specialists Conference,* 2004, vol. 6, pp: 4630–4636.
37. Celli, G.; Pilo, F.; and Tennakoon, S.B., "Voltage regulation of 25 kV AC railway systems by using thyristor switched capacitor", *Ninth International Conference on Harmonics and Quality of Power,* 2000, vol. 2, pp: 633–638.
38. Pee-Chin Tan; Poh Chiang Loh; and Holmes, D.G., "A robust multilevel hybrid compensation system for 25-kV electrified railway applications", *IEEE Transactions on Power Electronics,* 2004, vol. 19 issue: 4, pp: 1043–1052.
39. Uzuka, T.; Ikedo, S.; and Ueda, K., "A static voltage fluctuation compensator for AC electric railway", *IEEE 35th Annual Power Electronics Specialists Conference,* 2004, vol. 3, pp: 1869–1873.
40. Longhua Zhou; Qing Fu; Xiangfeng Li; and Changshu Liu, "A novel Multilevel Power Quality Compensator for electrified railway", *IEEE 6th International Power Electronics and Motion Control Conference,* 2009, pp: 1141–1147.
41. Wei Yingdong; and Han Yingduo, "Research on Control of Comprehensive Compensation for Traction Substations based on the STATCOM Technology", 2009.
42. Wang, H.; Liu, Y.; Yan, K.; Fu, Y.; Zhang, C., "Analysis of static VAr compensators installed in different positions in electric railways", *Electrical Systems in Transportation, IET*, 2015, vol: 5, issue: 3, pp: 129–134.
43. Dai, N.Y.; Lao, K.W.; Wong, M.C.; Wong, C.K., "Hybrid power quality conditioner for co-phase power supply system in electrified railway", *Power Electronics, IET*, 2012, vol: 5, issue: 7, pp: 1084–1094.
44. Keng-Weng Lao; NingYi Dai; WeiGang Liu; Man-Chung Wong; Chi-Kong Wong, "Modeling and control of Railway Static Power Conditioner compensation based on power quality standards", *2012 IEEE 13th Workshop on Control and Modeling for Power Electronics (COMPEL)*, 2012, pp: 1–6.

45. Lao, K.-W.; Dai, N.; Liu, W.-G.; Wong, M.-C., “Hybrid Power Quality Compensator With Minimum DC Operation Voltage Design for High-Speed Traction Power Systems”, *IEEE Transactions on Power Electronics*, 2013, vol: 28, issue: 4, pp: 2024–2036.
46. Keng-Weng Lao; Man-Chung Wong; NingYi Dai; Chi-Kong Wong; Chi-Seng Lam, “A Systematic Approach to Railway Hybrid Quality Conditioner Design With Harmonic Compensation for High-Speed Railway”, *IEEE Transactions on Industrial Electronics*, 2015, vol: 62, issue: 2, pp: 930–942.
47. Keng-Weng Lao; Man-Chung Wong; Ning Yi Dai; Chi-Kong Wong; Chi-Seng Lam, “Analysis of DC-Link Operation Voltage of a Hybrid Railway Power Quality Conditioner and Its PQ Compensation Capability in High-Speed Cophase Traction Power Supply”, IEEE Transactions on Power Electronics, 2016, vol: 31, issue: 2, pp: 1643–1656.
48. Keng-Weng Lao; Man-Chung Wong; NingYi Dai; Chi-Seng Lam; Chi-Kong Wong; Lei Wang, “Analysis in the Effect of Co-phase Traction Railway HPQC Coupled Impedance on Its Compensation Capability and Impedance-Mapping Design Technique Based on Required Compensation Capability for Reduction in Operation Voltage”, 2017, vol: 32, issue: 4, pp: 2631–2646.
49. Keng-Weng Lao; Man-Chung Wong; Ningyi Dai; Chi-Seng Lam; Lei Wang; Chi-Kong Wong, “Analysis of the Effects of Operation Voltage Range in Flexible DC Control on Railway HPQC Compensation Capability in High-Speed Co-phase Railway Power” IEEE Transactions on Power Electronics, 2018, vol: 33, issue: 2, pp: 1760–1774.
50. Ning Yi Dai; Keng-Weng Lao; Chi-Seng Lam, “Railway Hybrid Quality Conditioner With Partial Compensation for Converter Rating Reduction”, *IEEE Transactions on Industry Applications*, 2015, vol: 51, issue: 5, pp: 4130–4138.
51. Ning-Yi Dai; Man-Chung Wong; Keng-Weng Lao; Chi-Kong Wong, “Modelling and control of a railway power conditioner in co-phase traction power system under partial compensation”, *Power Electronics, IET*, 2014, vol: 7, issue: 5, pp: 1044–1054.
52. Akagi, Hirofumi; Kanazawa, Yoshihira; and Nabae, Akira, “Instantaneous Reactive Power Compensators Comprising Switching Devices without Energy Storage Components”, *IEEE Transactions on Industry Applications*, 1984, Vol. IA-20, issue: 3, pp: 625–630.
53. Kunshan Yu, “Electric Railway Power Supply and Power Quality”, China Electric Power Press, 2010.

Chapter 2
Co-phase Traction Power Supply with Railway HPQC: Modeling, Control, and Advantages Over System with RPC

Abstract In this chapter, the three major power quality problems, system unbalance, presence of reactive power, and harmonics, are modeled in co-phase traction power to develop the compensation power control of Railway HPQC. A systematic analysis is performed based on the control developed and the performance of co-phase traction power supply system with Railway HPQC is compared with that of RPC. The system of co-phase traction power supply with Railway HPQC is advantageous for lower operation voltage and device rating. These advantages are evaluated using a simulation model in PSCAD/EMTDC and are presented in this chapter.

2.1 Introduction

Although the impedance of Railway Hybrid Power Quality Conditioner (Railway HPQC) is capacitive, which is different from that of Railway Power Compensator (RPC) in [1–3], the required output compensation power remains the same since they are both connected across the two single-phase transformer outputs. The required output compensation power model acts as an important base for Railway HPQC studies [4–12]. However, most of the contents in the publication do not include much exploration on the discussions on the derivation of this power compensation. In this chapter, one method of deriving required compensation power for the power quality conditioner (either RPC or Railway HPQC) in co-phase traction power based on power quality model is presented. The circuit configuration and parameter definition are first introduced, followed by power quality modeling discussions. Finally, the required compensation power is derived based on the developed power quality model and the comprehensive control algorithm is presented. The final part of this chapter shows a comparison between RPC and Railway HPQC.

K.-W. Lao et al., *Co-phase Traction Power Supply with Railway Hybrid Power Quality Conditioner*, https://doi.org/10.1007/978-981-13-0438-5_2

2.2 System Configuration of Co-phase Traction Power Supply with Railway HPQC

2.2.1 Circuit Topology

The circuit topology of co-phase traction power supply system with Railway HPQC is shown in Fig. 2.1. The three-phase source power voltage is 110 or 220 kV. Three-phase source power is then transformed into two single phases, namely V_{ac} and V_{bc} phases, with 27.5 and 13.75 kV, respectively, via substation transformer. In this project, for simplicity, V/V transformer, which is composed of two simple single-phase transformers, is used. Locomotives loadings are usually electrified with 27.5 kV. All locomotive loadings are connected across the V_{ac} phase, with V_{bc} phase unloaded. The Railway HPQC is connected across V_{ac} and V_{bc} phases to provide power quality compensation from secondary to primary side. The dc operation voltage of the compensator Railway HPQC can then be reduced compared to traditional RPC. More will be discussed in later contents.

2.2.2 System Parameters

In order to do further analysis, the system parameters are defined, as shown in Figs. 2.2 and 2.3. Some of them are summarized in Tables 2.1 and 2.2.

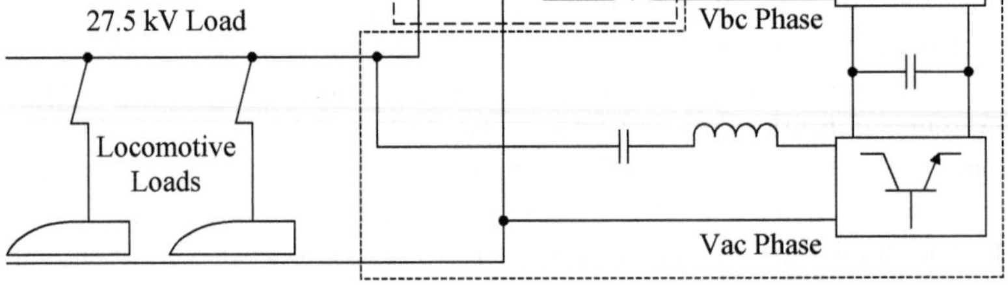

Fig. 2.1 Circuit schematic of co-phase traction power supply system with Railway HPQC

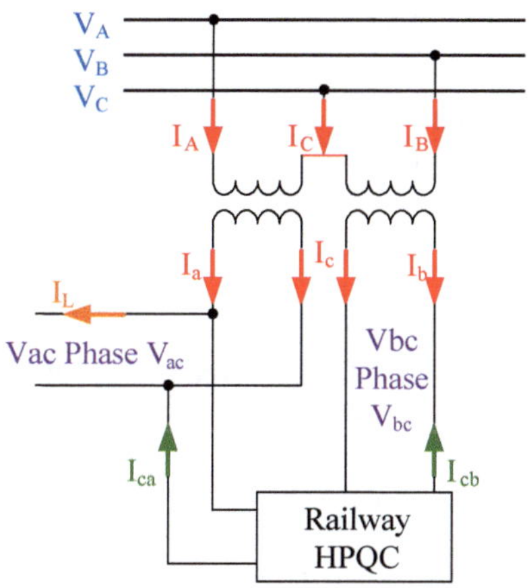

Fig. 2.2 Simplified model showing physical parameters of co-phase traction power supply with Railway HPQC

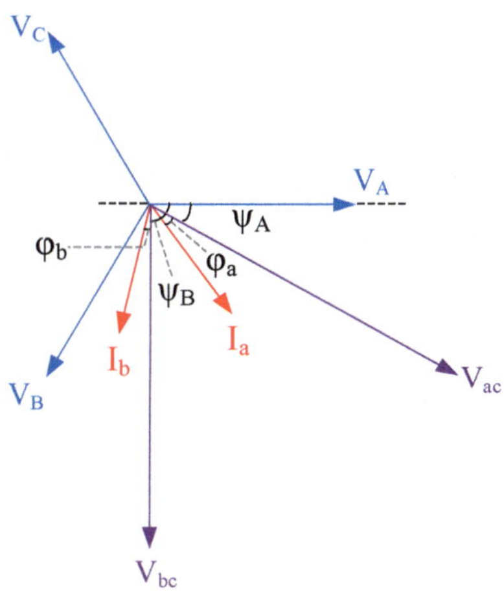

Fig. 2.3 Vector diagram showing the definition of phase angle in the co-phase traction power

Table 2.1 System parameter definition of co-phase traction power supply system (for modeling)

Items	Parameter(s)
Three-phase primary source voltage (rms value)	V_A, V_B, V_C
Three-phase primary source current (rms value)	I_A, I_B, I_C
V_{ac} phase secondary voltage and current (rms values)	V_{ac}, I_a
V_{bc} phase secondary voltage and current (rms values)	V_{bc}, I_b
Railway HPQC output compensation current to V_{ac} phase (rms)	I_{ca}
Railway HPQC output compensation current to V_{bc} phase (rms)	I_{cb}
Locomotive load current (rms value)	I_L

Table 2.2 Phase angle definition of the system

Items	Parameter(s)
Phase angle shift between V_A and V_{ac}	Ψ_A
Phase angle shift between V_B and V_{bc}	Ψ_B
Phase angle between V_{ac} and I_a	φ_a
Phase angle between V_{ac} and I_a	φ_b

Further parameter definition of different vectors in Fig. 2.3 is given in (2.1). It is useful for system modeling that follows:

$$\begin{cases} \dot{V}_{ac} = V_{ac}e^{-j\psi_A}, \quad \dot{I}_a = K\dot{I}_A = I_a e^{-j(\psi_A + \varphi_a)} \\ \dot{V}_{bc} = V_{bc}e^{-j\psi_B}, \quad \dot{I}_b = \left(\frac{K}{K_B}\right)\dot{I}_B = I_b e^{-j(\psi_B + \varphi_b)} \\ \qquad |V_{bc}| = K_B|V_{ac}| \end{cases} \tag{2.1}$$

2.3 Co-phase Traction Power Quality Problem Modeling

Since the usage of Railway HPQC in the co-phase system is for power quality compensation, the power quality problems should be modeled first. Based on the circuit topology and parameter definition in Sect. 2.1, the power quality problems in co-phase traction power supply can be modeled. More details are shown in this section.

In order to model the power quality problems, further assumptions are made. It is assumed that the system can be linearly decomposed into fundamental and harmonic model as shown in Fig. 2.4, in which system unbalance and reactive power are derived from fundamental one; while harmonics is derived from harmonic one.

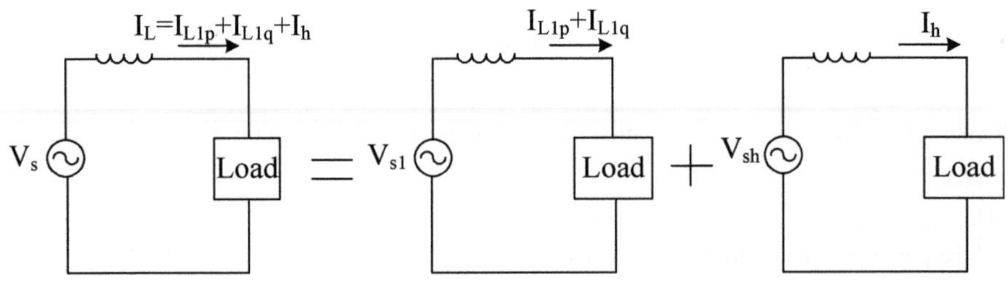

Fig. 2.4 It is assumed that the traction power system can be linearly decomposed into fundamental and harmonic models

2.3.1 System Unbalance and Negative Sequence Components

As introduced in Chap. 1, system unbalance in co-phase and other traction power system is mainly caused by the injection of negative sequence components from unbalanced locomotive loadings. The system unbalance in co-phase traction power supply is first modeled.

Referring to IEEE standard, system unbalance is defined as the ratio (%) of negative sequence current to positive sequence current in the system. Modeling of system unbalance is therefore equivalent to modeling of positive and negative sequence components. According to the method of symmetrical components, the zero, positive, and negative sequence components may be determined by (2.2), where F_{a0}, F_{a2}, and F_{a1} refer to zero(0), negative(−) and positive(+) sequence, respectively; while F_a, F_b, and F_c are the three-phase components.

$$\begin{bmatrix} \dot{F}_a \\ \dot{F}_b \\ \dot{F}_c \end{bmatrix} = \frac{1}{3}\begin{bmatrix} 1 & 1 & 1 \\ 1 & \alpha & \alpha^2 \\ 1 & \alpha^2 & \alpha \end{bmatrix}\begin{bmatrix} \dot{F}_{a0} \\ \dot{F}_{a2} \\ \dot{F}_{a1} \end{bmatrix} \tag{2.2}$$

where $\alpha = e^{+j\frac{2\pi}{3}} = e^{+j120^\circ} = -\frac{1}{2} + \frac{\sqrt{3}}{2}j$

Based on (2.2), the zero, negative, and positive sequence current component in co-phase traction power can be deduced as (2.3). Notice that all model must be expressed in terms of parameters in transformer secondary side. The value K, and K_B are added to model the transformer turn ratio, and the different turn ratio of V_{bc} phase transformer with V_{ac} one.

$$\begin{bmatrix} \dot{I}^0 \\ \dot{I}^- \\ \dot{I}^+ \end{bmatrix} = \frac{1}{3}\begin{bmatrix} 1 & 1 & 1 \\ 1 & \alpha^2 & \alpha \\ 1 & \alpha & \alpha^2 \end{bmatrix}\begin{bmatrix} \dot{I}_a \\ \dot{I}_b \\ \dot{I}_c \end{bmatrix} = \frac{1}{3K}\begin{bmatrix} 1 & 1 & 1 \\ 1 & \alpha^2 & \alpha \\ 1 & \alpha & \alpha^2 \end{bmatrix}\begin{bmatrix} \dot{I}_a \\ K_B\dot{I}_b \\ -\dot{I}_a - K_B\dot{I}_b \end{bmatrix} \tag{2.3}$$

where $\alpha = e^{+j\frac{2\pi}{3}} = e^{+j120^\circ} = -\frac{1}{2} + \frac{\sqrt{3}}{2}j$

By substituting (2.1) into (2.3), the expression in (2.4) can be obtained. For system unbalance, the useful information is the second one ($\dot{I}^-$). It will be further used to derive the system unbalance compensation algorithm in co-phase traction power.

$$\begin{cases} \dot{I}^0 = 0 \\ \dot{I}^- = \frac{1}{\sqrt{3}K}\left[I_a e^{-j(\psi_a+\varphi_a)} \cdot e^{-j30^\circ} + K_B I_b e^{-j(\psi_b+\varphi_b)} \cdot e^{-j90^\circ}\right] \\ \dot{I}^+ = \frac{1}{\sqrt{3}K}\left[I_a e^{-j(\psi_a+\varphi_a)} \cdot e^{+j30^\circ} + K_B I_b e^{-j(\psi_b+\varphi_b)} \cdot e^{+j90^\circ}\right] \end{cases} \tag{2.4}$$

2.3.2 System Source Reactive Power and Power Factor

Besides system unbalance, the presence of reactive power in the power source should also be avoided or at least minimized. In traction power system, reactive power mainly comes from inductive traction loading. Power factor is a common evaluation parameter for degree of presence of reactive power in system. Referring to Sect. 1.4.2.2 in Chap. 1, power factor is defined as the ratio of active Power P to apparent power S, in which $S^2 = P^2 + Q^2$ is satisfied in a harmonic free system. In other words, when the value of power factor is near 1.0, it indicates efficient usage of the power and less presence of reactive power.

In co-phase traction power supply, the presence of reactive power in primary source grid should be minimized. Therefore, the power factor at three-phase primary source grid is being modeled. Referring to Figs. 2.3 and 1.6, the expression in (2.5) can be obtained. It will be further used to derive the compensation for reactive power in co-phase traction power.

$$\begin{aligned} \mathrm{PFA} &= \cos(0^\circ + \psi_a + \phi_a) \\ \mathrm{PFB} &= \cos(-120^\circ + \psi_b + \phi_b) \\ \mathrm{PFC} &= \cos\left(-240^\circ + \tan^{-1}\left(\frac{(\frac{1}{K})I_a \sin\phi_a + (\frac{K_B}{K})I_b \sin\phi_b}{-(\frac{1}{K})I_a \cos\phi_a - (\frac{K_B}{K})I_b \cos\phi_b}\right)\right) \end{aligned} \tag{2.5}$$

where $\varphi_a = \psi_a + \phi_a$ and $\varphi_a = \psi_a + \phi_a$.

2.3.3 System Source Harmonics and Nonlinearity

Harmonic contents in the source grid of traction power supply system are mainly caused by the nonlinear characteristics of locomotive loadings. The system harmonics of co-phase traction power is modeled by using the harmonic model. It is further assumed that the harmonic model can be decomposed into n harmonic frequency models. In this case, the traction load harmonic contents could be represented by (2.6). During harmonic compensation, these harmonic contents are to be eliminated to guarantee that the system is harmonic free.

$$I_{\mathrm{Lh}} = \sqrt{\sum_{h=2}^{\infty} I_{\mathrm{Lh}\_h}^2} \tag{2.6}$$

2.4 Power Quality Compensation in Co-phase Traction Power

Details about power quality compensation algorithm in co-phase traction power are derived and presented in this section. The control algorithm works the same in conventional system with RPC and the new system with Railway HPQC. In the following analysis, the compensation algorithm is developed based on the composition of fundamental and harmonic compensation.

2.4.1 Fundamental Compensation: System Unbalance and Reactive Power

As discussed previously, fundamental compensation includes system unbalance and reactive power compensation. It is difficult to decompose them since compensation of system unbalance will affect reactive power, and vice versa. The expressions in (2.7) can be obtained by further combining and manipulating (2.4) and (2.5).

$$\begin{cases} \text{Unbalance: } \left|\dot{I}^{-}\right| = \frac{1}{\sqrt{3}K}\sqrt{I_a^2 + K_B^2 I_b^2 + 2I_a K_B I_b \cos(-120^\circ + \phi_a - \phi_b)} \\ \text{PFA} = \cos(\psi_{\text{A}} + \phi_a) \\ \text{PFB} = \cos(-120^\circ + \psi_{\text{B}} + \phi_b) \\ \text{PFC} = \cos\left(-240^\circ + \tan^{-1}\left(\frac{(1/\text{K})I_a \sin\varphi_a + (\text{K}_\text{B}/\text{K})I_b \sin\varphi_b}{-(1/\text{K})I_a \cos\varphi_a - (\text{K}_\text{B}/\text{K})I_b \cos\varphi_b}\right)\right) \end{cases} \tag{2.7}$$

where $\varphi_a = \psi_a + \phi_a$ and $\varphi_a = \psi_a + \phi_a$.

Furthermore, it is important that the power quality compensator, either RPC or Railway HPQC, does not consume or store any additional active power; otherwise, it would cause additional power loss. The relationship constraint is shown in (2.8).

$$I_a \cos\phi_a + I_b \cos\phi_b = I_L \cos\phi_L = I_{\text{Lp}} \tag{2.8}$$

By solving (2.7) and (2.8), the parameters for substation transformer secondary side after compensation can be obtained, as shown in (2.9).

$$\begin{cases} \phi_a = \cos^{-1}(\text{PFA}) - \psi_A \\ \phi_b = \cos^{-1}(\text{PFB}) + 120^\circ - \psi_B \\ I_a = -K_B I_b\left[\cos(-120^\circ + \phi_a - \phi_b) + \sqrt{(\cos(-120^\circ + \phi_a - \phi_b))^2 - 1}\right] \\ I_b = \frac{I_{Lp} - I_a \cos\phi_a}{\cos\phi_b} \end{cases} \tag{2.9}$$

With reference to Fig. 2.2, the relationship between Railway HPQC compensation current I_{ca} and I_{cb} with substation transformer secondary side current can be concluded through simple circuit analysis, as shown in (2.10).

$$\begin{aligned}\dot{I_{ca}} &= \dot{I_a} - \dot{I_L} = I_a e^{-j(\psi_A + \phi_a)} - I_L e^{-j(\psi_A + \phi_L)} \\ \dot{I_{cb}} &= -\dot{I_b} = -I_b e^{-j(-120^\circ + \psi_B - \phi_b)}\end{aligned} \tag{2.10}$$

It has been emphasized that power quality compensation in co-phase traction power supply is provided from secondary side to primary side. Therefore, the expression in (2.10) is revised such that the vector zero angle reference is modified as V_{ac} and V_{bc} respectively, as shown in (2.11).

$$\begin{aligned}\dot{I_{ca}} &= I_L e^{-j(\phi_L)} - I_a e^{-j(\phi_a)} = I_L(\cos\phi_L - j\sin\phi_L) - I_a(\cos\phi_a - j\sin\phi_a) \\ \dot{I_{cb}} &= -I_b e^{-j(-\phi_b)} = -I_b(\cos\phi_b - j\sin\phi_b)\end{aligned} \tag{2.11}$$

(2.11) can then be further simplified in terms of load active and reactive power I_{Lp} and I_{Lq}, and is shown in (2.12).

$$\begin{aligned}\dot{I_{ca}} &= I_{ca} e^{-j\theta_{ca}} = \dot{I_{cap}} + \dot{I_{caq}} = (I_{Lp} - I_a\cos\phi_a) - j(I_{Lq} - I_a\sin\phi_a) \\ \dot{I_{cb}} &= I_{cb} e^{-j\theta_{cb}} = \dot{I_{cbp}} + \dot{I_{cbq}} = -I_b e^{-j(-\phi_b)} = -I_b(\cos\phi_b - j\sin\phi_b)\end{aligned} \tag{2.12}$$

Two new parameters θ_{ca} and θ_{cb} are added to represent the power angle of the compensation current I_{ca} and I_{cb}, with V_{ac} and V_{bc}, respectively. According to (2.12), they may be determined by (2.13)

$$\begin{aligned}\theta_{ca} &= \tan^{-1}\left(\frac{I_{Lq} - I_a\sin\phi_a}{I_{Lp} - I_a\cos\phi_a}\right) \\ \theta_{cb} &= \tan^{-1}\left(\frac{-\sin\phi_b}{-\cos\phi_b}\right)\end{aligned} \tag{2.13}$$

Referring to (2.9) and (2.12), it can be observed that the values of I_{cap}, I_{cbp}, and I_{cbq} can be expressed in terms of I_{Lp}; while the value of I_{caq} is expressed of both I_{Lq} and I_{Lq}. The required compensation current can thus be derived as shown in (2.14).

$$\begin{bmatrix} I_{cap} \\ I_{caq} \\ I_{cbp} \\ I_{cbq} \end{bmatrix} = \begin{bmatrix} K_1 I_{Lp} \\ K_2 I_{Lp} + I_{Lq} \\ K_3 I_{Lp} \\ K_4 I_{Lp} \end{bmatrix} \tag{2.14}$$

In power quality compensation, the instantaneous pq theory is commonly used. The compensation algorithm is usually expressed in terms of power. Therefore, the final control algorithm is expressed in terms of load active (p_L) and reactive power

(q_L), as shown in (2.15). The values of K_1, K_2, K_3, and K_4 are coefficients and are to be determined according to the power quality requirement.

$$\begin{bmatrix} p_{caf} \\ q_{caf} \\ p_{cbf} \\ q_{cbf} \end{bmatrix} = \begin{bmatrix} K_1\bar{p}_L \\ K_2\bar{p}_L + \bar{q}_L \\ K_3\bar{p}_L \\ K_4\bar{p}_L \end{bmatrix} \quad (2.15)$$

Example

For easier understanding, two examples are shown here for reference. It is assumed that the power quality requirement is selected as:

(a) System Unbalance = 0%, PFA = PFB = 1.0 (ideal case: zero unbalance and reactive power);
(b) System Unbalance = 0%, PFA = PFB = 0.89 (zero unbalance, reactive power = half of active power)

It is assumed that the V/V transformer connected is used such that $\Psi_A = 30°$, $\Psi_B = 90°$. The values of I_a, I_b, φ_a, and φ_b can be computed by substituting the requirement above into (2.9). The values of K_1, K_2, K_3 and K_4 can then be computed based on the values of I_{ca} and I_{cb} in (2.12). The computed parameters are shown in Table 2.3.

Table 2.3 Example computations of power quality compensation control algorithm coefficients according to the compensation requirement

	Condition A	Condition B
Parameters	$\|I^-\| = 0$, PFA = PFB = 1.0	$\|I^-\| = 0$, PFA = PFB = 0.89
V_{ac} phase current, I_a	$0.5774I_{Lp}$	$0.6479I_{Lp}$
V_{bc} phase current, I_b	$0.5774I_{Lp}$	$0.6479I_{Lp}$
V_{ac} current angle, φ_a	−30°	−2.87°
V_{bc} current angle, φ_b	+30°	57.13°
V_{ac} compensation active current, I_{cap}	$0.5I_{Lp}$	$0.3508I_{Lp}$
V_{ac} compensation reactive current, I_{caq}	$0.2887I_{Lp} + I_{Lq}$	$0.0325I_{Lp} + I_{Lq}$
V_{bc} compensation active current, I_{cbp}	$-0.5I_{Lp}$	$-0.3508I_{Lp}$
V_{bc} compensation reactive current, I_{cbq}	$-0.2887I_{Lp}$	$-0.0325I_{Lp}$
Coefficient K_1	0.5	0.3508
Coefficient K_2	0.2887	0.0325
Coefficient K_3	−0.5	−0.3508
Coefficient K_4	−0.2887	−0.5459

2.4.2 Harmonic Compensation: System Source Harmonics

Next, the harmonic compensation is derived based on the harmonic model. Referring back to the model in (2.6), the harmonic contents mainly come from locomotive load and are composed of components at different frequencies except fundamental one. In the compensation theory developed in instantaneous pq theory, these components refer to the oscillating portions of instantaneous active and reactive power. Since harmonics mainly come from the locomotive load at V_{ac} phase, there is no harmonic compensation requirement at V_{bc} phase, the harmonic compensation algorithm is therefore shown as in (2.16).

$$\begin{bmatrix} p_{cah} \\ q_{cah} \\ p_{cbh} \\ q_{cbh} \end{bmatrix} = \begin{bmatrix} \tilde{p}_L \\ \tilde{q}_L \\ 0 \\ 0 \end{bmatrix} \tag{2.16}$$

2.4.3 Comprehensive Compensation Control Algorithm

Based on the derivation in Sects. 2.3.1 and 2.3.2, the comprehensive power quality compensation control algorithm can then be derived by the linear composition of the fundamental and harmonic compensation control algorithm in (2.15) and (2.16). The final comprehensive control algorithm is shown in (2.17).

$$\begin{bmatrix} p_{ca} \\ q_{ca} \\ p_{cb} \\ q_{cb} \end{bmatrix} = \begin{bmatrix} p_{caf} + p_{cah} \\ q_{caf} + q_{cah} \\ p_{cbf} + p_{cbh} \\ q_{cbf} + q_{cbh} \end{bmatrix} = \begin{bmatrix} K_1\bar{p}_L + \tilde{p}_L \\ K_2\bar{p}_L + \bar{q}_L + \tilde{q}_L \\ K_3\bar{p}_L \\ K_4\bar{p}_L \end{bmatrix} = \begin{bmatrix} K_1\bar{p}_L + \tilde{p}_L \\ K_2\bar{p}_L + q_L \\ K_3\bar{p}_L \\ K_4\bar{p}_L \end{bmatrix} \tag{2.17}$$

Example

Taking the same example as in previous example, the comprehensive control algorithm under two different power quality requirement condition is shown in Table 2.4. Notice that the positive sign of V_{ac} phase compensation current refers to

Table 2.4 Comprehensive power quality compensation control algorithm in co-phase traction power according to different requirement

	Condition A	Condition B
	$\lvert I^{-}\rvert = 0$, PFA = PFB = 1.0	$\lvert I^{-}\rvert = 0$, PFA = PFB = 0.89
Comprehensive control algorithm	$\begin{bmatrix} p_{ca} \\ q_{ca} \\ p_{cb} \\ q_{cb} \end{bmatrix} == \begin{bmatrix} 0.5\bar{p}_L + \tilde{p}_L \\ 0.2887\bar{p}_L + q_L \\ -0.5\bar{p}_L \\ -0.2887\bar{p}_L \end{bmatrix}$	$\begin{bmatrix} p_{ca} \\ q_{ca} \\ p_{cb} \\ q_{cb} \end{bmatrix} == \begin{bmatrix} 0.3508\bar{p}_L + \tilde{p}_L \\ 0.0325\bar{p}_L + q_L \\ -0.3508\bar{p}_L \\ -0.5459\bar{p}_L \end{bmatrix}$

compensation power emission to the system, while the negative sign of V_{bc} phase compensation current refers to power absorption from the system.

2.4.4 Further Analysis of Co-phase Power Quality Compensation Algorithm

During the exploration of compensation control algorithm in co-phase traction power, there are some interesting findings that worth attention, but does not affect the derivation. They are being categorized and discussed in this subsection.

Phasor diagram showing operations with and without compensation

In order to have deeper investigations, phase diagrams showing the operations with and without power quality compensation in co-phase traction power supply are constructed. Shown in Fig. 2.5 is the phasor diagram of co-phase traction power supply without any compensation. It can be observed that the system source current is greatly unbalanced (notice that the current I_B is zero). Moreover, there is a large angle between source voltage and current, which indicates the presence of reactive power. The power quality performance in co-phase traction power when no compensation is provided is far from satisfactory.

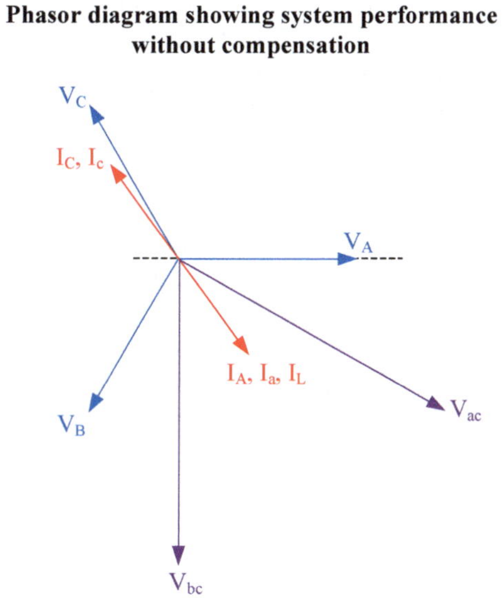

Fig. 2.5 Phasor diagram showing system performance of co-phase traction power supply without any compensation

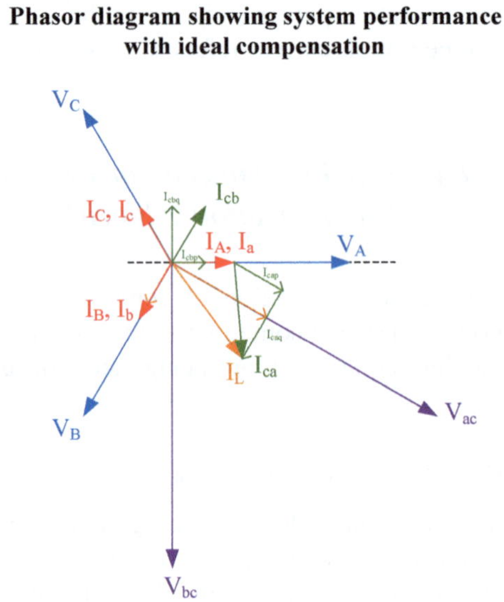

Fig. 2.6 Phasor diagram showing system performance of co-phase traction power supply with ideal power quality compensation

Next, the condition which ideal power quality compensation is provided is investigated. In all contents that follow, ideal power quality compensation refers to the power quality requirement of zero system unbalance, unity power factor, and zero harmonic current. The phase diagram under ideal compensation is shown in Fig. 2.6.

The figure is constructed in proportional and ratio according to the derivations in (2.17) and Table 2.4. It can be observed from the figure that after ideal compensation, the system source current is balanced (notice that amplitude of I_A, I_B and I_C are the same, with 120° phase difference between one another); the reactive power in the source grid is also eliminated (notice that the angle differences between V_A and I_A, V_B and I_B, V_C and I_C, are all zeros).

Referring back to (2.17) and Table 2.4, it is noticed that the reactive compensation power does not include only load reactive power, but also a certain additional component in proportion with load active power By further investigating Fig. 2.6, it is also interesting to find that the additional reactive compensation power is in some sense used to compensate the phase shift between primary (V_A, V_B, V_C) and secondary (V_{ac}, V_{bc}) of substation transformer. This makes the co-phase traction power quality compensator different from others.

Relationship between reactive power and system unbalance

It has already been mentioned previously that it is rather difficult to decouple the effect of reactive power and system unbalance. However, it is found that when the value of PFA and PFB is not the same, the negative sequence component cannot be zero. In other words, the system can never be balanced.

The proof goes as follows. Referring back to (2.9), the value of φ_a and φ_b is mainly dependent on the value of PFA and PFB. Furthermore, for the expression of I_a to have rational solution at zero unbalance condition, the condition in (2.18) must be satisfied.

$$(\cos(-120^\circ + \phi_a - \phi_b))^2 \geq 1 \tag{2.18}$$

For a cosine square function, it is impossible to have value larger than 1. Therefore, there is only one condition left, that is, when the expression in (2.18) equals to 1. Under this condition, $\varphi_a - \varphi_b = -60^\circ$. In co-phase traction with V/V transformer, $\Psi_A = 30^\circ$, $\Psi_B = 90^\circ$, it can be concluded that the condition can be satisfied only when PFA = PFB. Therefore, the system can never be balanced when the value of PFA is not equal to that of PFB.

2.4.4.1 Active and Reactive Power Transfer

It can be concluded from (2.17) that different from traditional power quality compensator which usually involves reactive power compensation only, power quality compensation in co-phase traction power refers to both active and reactive power transfer.

As introduced, no active power is stored in the power quality compensator, such that the amount of active power absorbed from V_{bc} phase is always equal to the active power injected from V_{ac} phase to the system. In other words, $K_3 = -K_1$.

In contrast, for reactive power transfer, the reactive power absorbed from V_{bc} phase equals reactive power emitted from V_{ac} phase only under ideal compensation requirement (zero unbalance, unity power factor).

2.5 Co-phase Traction Power Quality Compensation Control Block Diagram

The power quality compensation control block diagram of co-phase traction power is shown in Fig. 2.7. Each block is introduced in details below.

2.5.1 Single-Phase Instantaneous Pq Computations

The load voltage (v_{ac}) and current (i_{La}) is first obtained from transducer and signal conditioning board. They are then being used to compute the load instantaneous active and reactive power based on the instantaneous active and reactive (instantaneous pq) theory developed by Prof. Akagi [1.52]. As traditional instantaneous pq

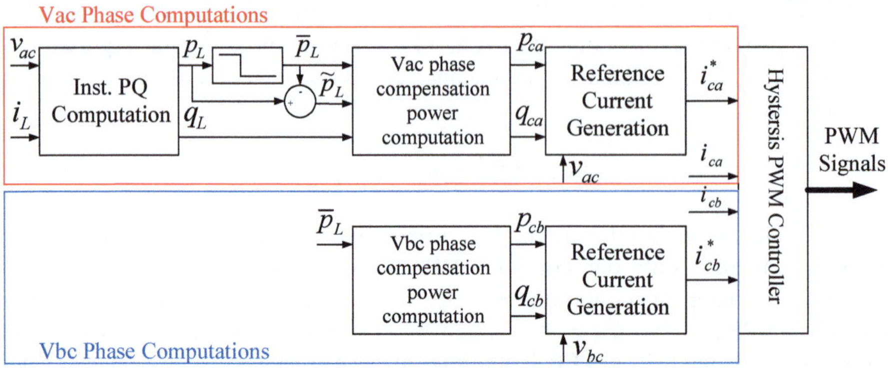

Fig. 2.7 Control block diagram of compensation control algorithm in co-phase traction power supply

theory is developed for three-phase power, some modifications are done to make it suitable for single-phase computation like the one in co-phase traction power supply.

First of all, it is assumed that the single-phase system is duplicated three times such that it forms a balanced three-phase circuit, with 120° phase difference between one another. Since this three-phase system does not exist in physical form, it is regarded as virtual one, with voltage and current definition shown in (2.19). Although three-phase system concept is used, mainly virtual A phase data is being used. Therefore, it does not affect the single-phase compensation.

$$\begin{aligned}
&\text{virtual three phase voltage:} \begin{cases} v_{\mathrm{acA}} = \sqrt{2}V_{\mathrm{ac}}\sin(\omega t) \\ v_{\mathrm{acB}} = \sqrt{2}V_{\mathrm{ac}}\sin(\omega t - 120°) \\ v_{\mathrm{acC}} = \sqrt{2}V_{\mathrm{ac}}\sin(\omega t + 120°) \end{cases} \\
&\text{virtual three phase current:} \begin{cases} i_{\mathrm{LA}} = \sqrt{2}I_L\sin(\omega t - \phi_L) \\ i_{\mathrm{LB}} = \sqrt{2}I_L\sin(\omega t - \phi_L - 120°) \\ i_{\mathrm{LC}} = \sqrt{2}I_L\sin(\omega t - \phi_L + 120°) \end{cases}
\end{aligned} \tag{2.19}$$

In the instantaneous pq theory, the Clarke transform is used. The three-phase coordinate system is transformed into an orthogonal system with zero, alpha, and beta components, using the transformation matrix shown in (2.20) and (2.21).

$$\begin{bmatrix} v_0 \\ v_\alpha \\ v_\beta \end{bmatrix} = \sqrt{\frac{2}{3}} \begin{bmatrix} \frac{1}{\sqrt{2}} & \frac{1}{\sqrt{2}} & \frac{1}{\sqrt{2}} \\ 1 & -\frac{1}{2} & -\frac{1}{2} \\ 0 & \frac{\sqrt{3}}{2} & -\frac{\sqrt{3}}{2} \end{bmatrix} \begin{bmatrix} v_{\mathrm{acA}} \\ v_{\mathrm{acB}} \\ v_{\mathrm{acC}} \end{bmatrix} \tag{2.20}$$

$$\begin{bmatrix} i_0 \\ i_\alpha \\ i_\beta \end{bmatrix} = \sqrt{\frac{2}{3}} \begin{bmatrix} \frac{1}{\sqrt{2}} & \frac{1}{\sqrt{2}} & \frac{1}{\sqrt{2}} \\ 1 & -\frac{1}{2} & -\frac{1}{2} \\ 0 & \frac{\sqrt{3}}{2} & -\frac{\sqrt{3}}{2} \end{bmatrix} \begin{bmatrix} i_{LA} \\ i_{LB} \\ i_{LC} \end{bmatrix} \tag{2.21}$$

By substituting (2.19) into (2.20) and (2.21), the zero, alpha, and beta components can be obtained as shown in (2.22) and (2.23).

$$\begin{bmatrix} v_0 \\ v_\alpha \\ v_\beta \end{bmatrix} = \begin{bmatrix} 0 \\ \sqrt{3} v_{ac} \sin(\omega t) \\ \sqrt{3} v_{ac} \cos(\omega t) \end{bmatrix} \tag{2.22}$$

$$\begin{bmatrix} i_0 \\ i_\alpha \\ i_\beta \end{bmatrix} = \begin{bmatrix} 0 \\ \sqrt{3} i_L \sin(\omega t - \phi_L) \\ \sqrt{3} i_L \cos(\omega t - \phi_L) \end{bmatrix} \tag{2.23}$$

It can be observed that with co-phase traction power configuration, the zero component is always zero, while the beta component is always lagging alpha component by 90°. Therefore, it is not necessary to perform three-phase instantaneous pq theory in the control block. Instead, the alpha and beta phase component may be determined by (2.24) and (2.25). The coefficient is removed since it will be canceled in later computations and can be ignored here.

$$\begin{bmatrix} v_\alpha \\ v_\beta \end{bmatrix} = \begin{bmatrix} v_{ac} \\ v_{ac} e^{-j\pi/2} \end{bmatrix} \tag{2.24}$$

$$\begin{bmatrix} i_\alpha \\ i_\beta \end{bmatrix} = \begin{bmatrix} i_L \\ i_L e^{-j\pi/2} \end{bmatrix} \tag{2.25}$$

The instantaneous load active and reactive power can then be determined by ordinary instantaneous pq theory, as shown in (2.26). The load power data can then be used to compute the required compensation power.

$$\begin{bmatrix} p_L \\ q_L \end{bmatrix} = \begin{bmatrix} v_\alpha & v_\beta \\ v_\beta & -v_\alpha \end{bmatrix} \begin{bmatrix} i_\alpha \\ i_\beta \end{bmatrix} = \begin{bmatrix} v_{ac} & v_{ac} e^{-j\pi/2} \\ v_{ac} e^{-j\pi/2} & -v_{ac} \end{bmatrix} \begin{bmatrix} i_L \\ i_L e^{-j\pi/2} \end{bmatrix} \tag{2.26}$$

2.5.2 *Computation of Required Active and Reactive Compensation Power*

After the computation of instantaneous load active and reactive power, the required active and reactive compensation power can then be determined according to (2.17) and the power quality requirement. As discussed and can be observed from the algorithm, it is required to differentiate the constant and oscillating portions in the instantaneous active power. This is done so using a low-pass filter. The filtered result refers to the constant portion; while the oscillating portion is obtained by subtracting the constant portion from the whole.

2.5.3 Determination of Required Compensation Current

Although the compensation control algorithm is derived based on the compensation power, the ultimate goal of the control block is to generate PWM signals to control the electronic switches in the power quality compensator and output required compensation current. Thus, it is important to derive the compensation current reference. This is done so by performing the inverse Clarke transform, as shown in (2.27).

$$\begin{cases} i_{\mathrm{ca}}^{*} = \dfrac{1}{v_{\mathrm{ac}}^{2} + \left(v_{\mathrm{ac}}e^{-j\pi/2}\right)^{2}} \begin{bmatrix} v_{\mathrm{ac}} & v_{\mathrm{ac}}e^{-j\pi/2} \end{bmatrix} \begin{bmatrix} p_{\mathrm{ca}} \\ q_{\mathrm{ca}} \end{bmatrix} \\ i_{\mathrm{cb}}^{*} = \dfrac{1}{v_{\mathrm{bc}}^{2} + \left(v_{\mathrm{bc}}e^{-j\pi/2}\right)^{2}} \begin{bmatrix} v_{\mathrm{bc}} & v_{\mathrm{bc}}e^{-j\pi/2} \end{bmatrix} \begin{bmatrix} p_{\mathrm{cb}} \\ q_{\mathrm{cb}} \end{bmatrix} \end{cases} \quad (2.27)$$

2.5.4 Hysteresis PWM Controller to Generate PWM Signals

Obtaining the current reference signal in (2.27) is still not enough for generating the compensation current. As introduced, the electronic switches within the power quality compensator should be appropriately controlled by PWM signals to control the power flow and compensation current. The most straightforward PWM control method is by using hysteresis method. An illustrative diagram is shown in Fig. 2.8. Usually, a hysteresis band (HB) is set based on the harmonic distortions tolerance. At each computation cycle, the actual compensation current is compared with the reference signal based on (2.28), the converter can then "track" the current reference and output the required compensation current

$$\begin{cases} \text{when } i_{\mathrm{ca}} - i*_{\mathrm{ca}} \geq \mathrm{HB}, \text{PWM control to generate lower } i_{\mathrm{ca}} \\ \text{when } i_{\mathrm{ca}} - i*_{\mathrm{ca}} \leq -\mathrm{HB}, \text{PWM control to generate higher } i_{\mathrm{ca}} \end{cases} \quad (2.28)$$

In the figure, the actual compensation current is represented by the thick saw-wave line. The saw wave represents the switching mechanism of (2.28); while the dotted lines represent the boundary conditions. Notice that the absolute value of

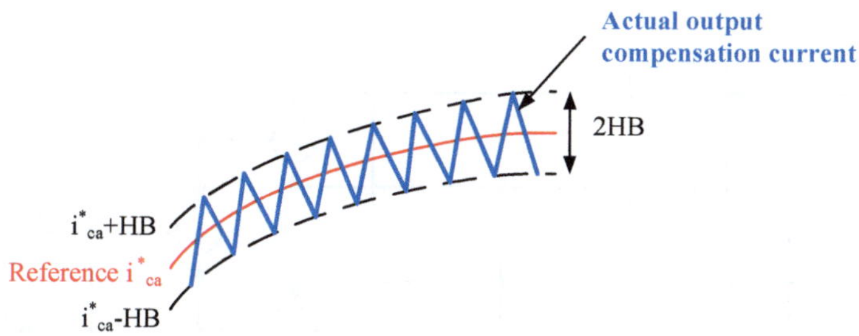

Fig. 2.8 Illustrative diagram of the working principle in hysteresis PWM control

the error between the actual output compensation current and the reference current is no larger than HB.

Based on the discussions above, the comprehensive control block diagram showing detailed computation algorithm is shown in Fig. 2.9.

2.6 Co-phase Traction Power System with Different Compensation Devices

As introduced, Railway HPQC is being used and implemented in the system. Conventionally, the traditional RPC is being used but its operation voltage is high. Therefore, the coupling structure and characteristics are modified to provide additional voltage support and lower down its operation voltage. More details about the difference between conventional system with RPC and new one with Railway HPQC, as well as the analysis based on the control algorithm above, will be presented in the following two sections.

2.6.1 Conventional System with Inductive-Coupled RPC

In conventional co-phase traction power supply system, inductive-coupled RPC is used to provide power quality compensation, as shown in Fig. 2.10. Comparing it with Fig. 2.1, it can be observed that in conventional RPC, the coupled impedance of the V_{ac} phase converter is inductive, and causes high voltage drop across it during compensation. More details about operation voltage requirement will be discussed later. The system parameter definitions are also shown in Fig. 2.11 for further analysis in the next section.

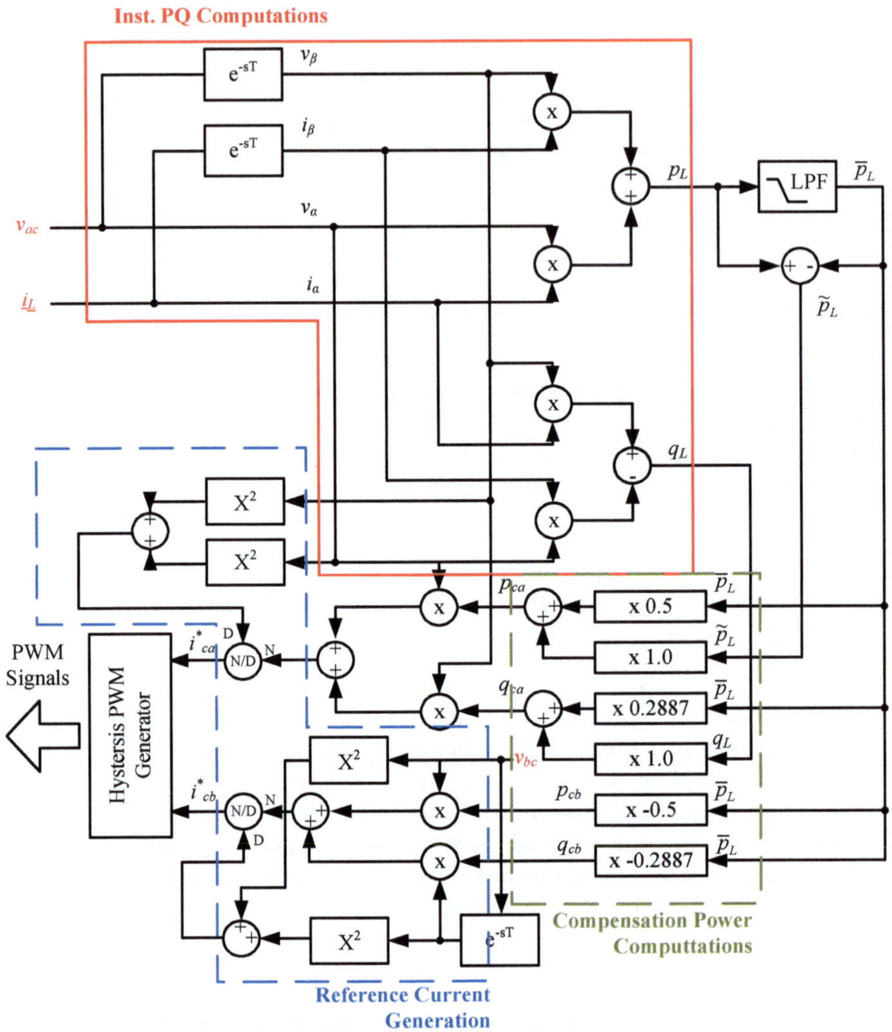

Fig. 2.9 Detailed control block diagram of power quality compensator in co-phase traction power

2.6.2 *Novel System with Capacitive-Coupled HPQC*

As shown in Fig. 2.1, different from conventional co-phase traction power supply, the capacitive-coupled Railway HPQC is being used to provide power quality compensation in the system. The figure is repeated in Fig. 2.12 for reference. Compared to conventional system with RPC in Fig. 2.10, the major difference is the coupling structure used in Railway HPQC V_{ac} phase converter. The coupling structure used is a series inductor–capacitor (LC) branch. It will be discussed in

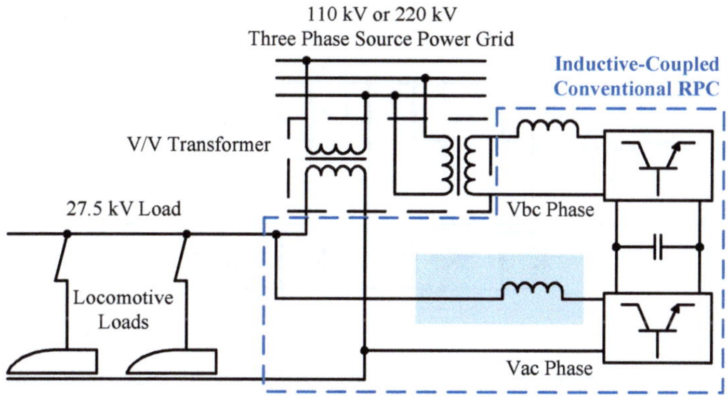

Fig. 2.10 Conventional co-phase traction power supply system configuration with inductive-coupled RPC

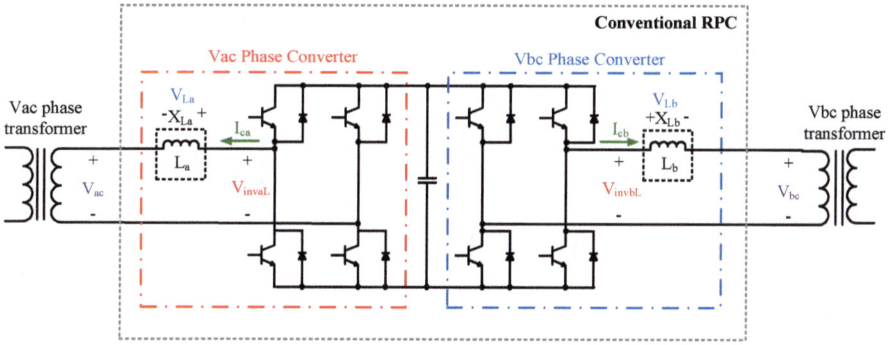

Fig. 2.11 System parameter definition of RPC in conventional co-phase traction power supply system for further analysis

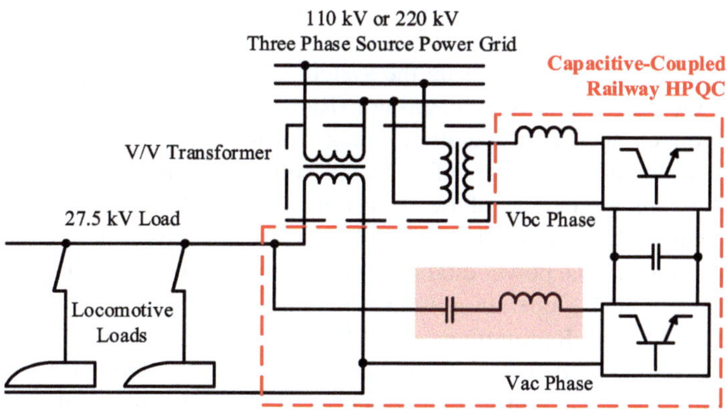

Fig. 2.12 System configuration of co-phase traction power supply with capacitive-coupled Railway HPQC

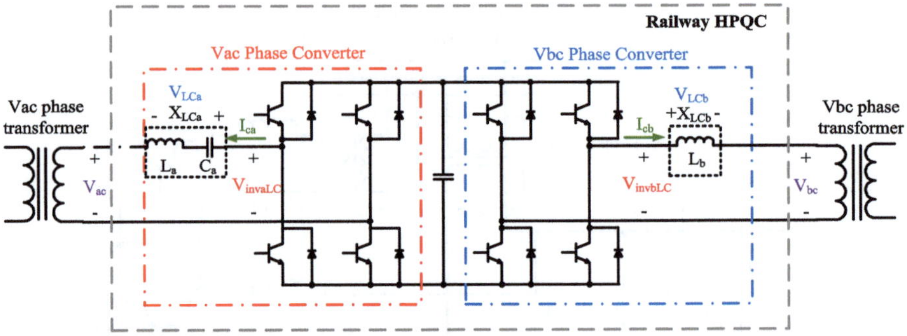

Fig. 2.13 System parameter definition of Railway HPQC in co-phase traction power supply system for further analysis

later section for the application of this series LC branch structure in reducing the Railway HPQC operation voltage. Similarly, the system parameter definition of Railway HPQC is shown in Fig. 2.13 for reference in the following analysis.

2.7 System Analysis and Comparisons Between Conventional RPC and Novel Railway HPQC

In this section, system analysis is performed on conventional RPC and novel Railway HPQC, comparisons are made between them to show the advantages of reduced operation voltage reduction in the system with Railway HPQC.

2.7.1 Reduction in Operation Voltage and Inverter Capacity Rating

As has been emphasizing, by using the system with Railway HPQC, the operation voltage of the power quality compensator (Railway HPQC) can be reduced. In the following, the compensator operation voltage for RPC and Railway HPQC will be derived. The operation voltage refers to the voltage requirement of the shared dc link of the V_{ac} phase and V_{bc} phase converter, which is determined by the required inverter voltage V_{inv}. Notice that in the following analysis, it is assumed that the operation voltage of the power quality compensator, either RPC or Railway HPQC, is dominant by the V_{ac} operation voltage requirement.

Operation Voltage Requirement of RPC

First, in order to deduce the operation voltage requirement of RPC, it is important to understand that when current passes through passive element, voltage potential different will be resulted. In co-phase traction power, the compensation current output by power quality compensator like RPC must follow the control algorithm derived in (2.17) and Table 2.4.

Based on this, the phasor diagram showing the operation condition of V_{ac} phase converter in conventional RPC during co-phase traction power quality compensation is shown in Fig. 2.14.

Referring back to (2.17) and Table 2.4, it can be observed that during compensation, reactive power is required to be output by the V_{ac} phase converter. This explains why a lagging inductive compensation current I_{ca} is shown in Fig. 2.14. It is composed of active and reactive component, I_{cap} and I_{caq}, as presented in (2.29).

$$\vec{I}_{ca} = |I_{cap}| - j|I_{caq}| \tag{2.29}$$

When an inductive compensation current passes through the coupled inductance L_a, it results in a voltage drop across the coupled inductor, and its amplitude is determined by (2.30).

$$\vec{V}_{La} = \vec{I}_{ca} \cdot \vec{X}_{La} \tag{2.30}$$

The expression in (2.31 is obtained by substituting (2.29) into (2.30)

$$\vec{V}_{La} = \left(|I_{cap}| - j|I_{caq}|\right) \cdot \vec{X}_{La} \tag{2.31}$$

It can also be observed from Fig. 2.14 that the RPC inverter output voltage, V_{invaL}, is the addition of PCC voltage vector, V_{ac}, and voltage across coupled impedance, V_{La}, as shown in (2.32).

$$\vec{V}_{invaL} = \vec{V}_{ac} + \vec{V}_{La} = \vec{V}_{ac} + \vec{I}_{ca} \cdot \vec{X}_{La} \tag{2.32}$$

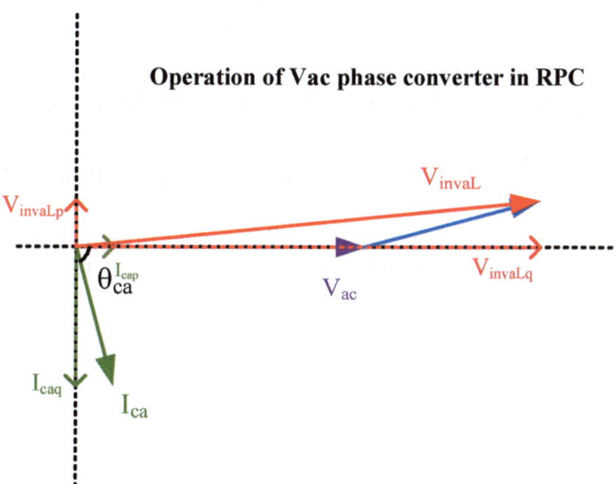

Fig. 2.14 Phase diagram showing the operation of V_{ac} phase converter in conventional RPC during compensation

With the coupled impedance X_{La} in conventional RPC being inductive, by substituting the relationship in (2.31), the expression in (2.32) is revised as in (2.33).

$$\vec{V}_{invaL} = \vec{V}_{ac} + \vec{I}_{ca} \cdot \vec{X}_{La} = V_{ac} + \left(I_{cap} - jI_{caq}\right) \cdot j|X_{La}| \tag{2.33}$$

The active and reactive components of V_{invaL} may then be further decomposed from (2.33), as shown in (2.34).

$$\vec{V}_{invaL} = V_{invaLp} - jV_{invaLq} = \left(V_{ac} + I_{caq} \cdot |X_{La}|\right) - j\left(-I_{cap} \cdot |X_{La}|\right) \tag{2.34}$$

The amplitude of V_{invaL} in conventional RPC can then be deduced as in (2.35).

$$\left|\vec{V}_{invaL}\right| = \sqrt{\left(V_{invaLp}\right)^2 + \left(V_{invaLq}\right)^2} = \sqrt{\left(V_{ac} + I_{caq} \cdot |X_{La}|\right)^2 + \left(I_{cap} \cdot |X_{La}|\right)^2} \tag{2.35}$$

With all parameters as positive element and all additional operations in (2.35), it can be inferred that the operation voltage rating V_{invaL} is always larger than the value of PCC voltage V_{ac}. The high operation voltage requirement is caused by the inductive current passing through an inductor. Therefore, the inductive characteristics of the coupled impedance in RPC leads to high operation voltage requirement.

2.7.1.1 Operation Voltage Requirement of Railway HPQC

On the other hand, the operation voltage requirement of Railway HPQC can be lower than the PCC operation voltage V_{ac}. More details about the derivation and analysis are provided below.

Shown in Fig. 2.15 is a phasor diagram showing the operation of V_{ac} phase converter in Railway HPQC during compensation. Similarly, the compensation current output requirement follows those in (2.17) and Table 2.4, and can be decomposed into active and reactive components.

By doing similar analysis as in RPC, the voltage across the capacitive-coupled impedance in Railway HPQC, V_{LCa}, is determined by (2.36).

$$\vec{V}_{LCa} = \left(\left|I_{cap}\right| - j\left|I_{caq}\right|\right) \cdot \vec{X}_{LCa} \tag{2.36}$$

Similar to conventional RPC, the inverter operation voltage V_{invaLC} can be computed by the addition of vector V_{ac} and V_{invaLC}, as shown in (2.37).

$$\vec{V}_{invaLC} = \vec{V}_{ac} + \vec{V}_{LCa} = \vec{V}_{ac} + \vec{I}_{ca} \cdot \vec{X}_{LCa} \tag{2.37}$$

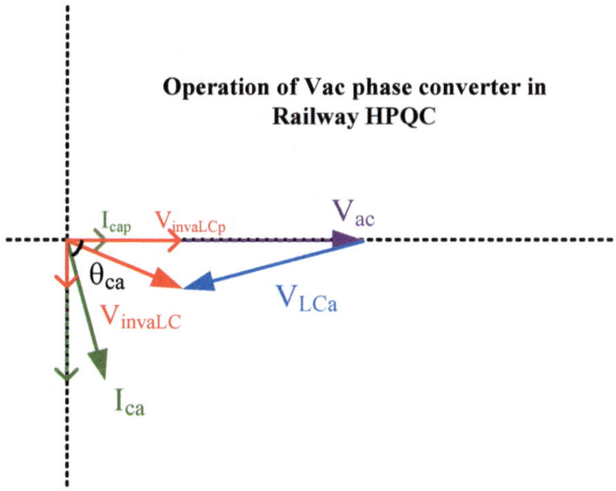

Fig. 2.15 Phase diagram showing the operation of V_{ac} phase converter in Railway HPQC during compensation

Since the coupled impedance X_{LCa} is capacitive in Railway HPQC, the expression in (2.37) can be revised as (2.38)

$$\vec{V}_{invaLC} = \vec{V}_{ac} + \vec{I}_{ca} \cdot \vec{X}_{LCa} = V_{ac} + \left(I_{cap} - jI_{caq}\right) \cdot -j|X_{LCa}| \tag{2.38}$$

The active and reactive components of V_{invaL} may then be further decomposed from (2.38), as shown in (2.39).

$$\vec{V}_{invaLC} = V_{invaLCp} - jV_{invaLCq} = \left(V_{ac} - I_{caq} \cdot |X_{LCa}|\right) - j\left(I_{cap} \cdot |X_{LCa}|\right) \tag{2.39}$$

The amplitude of V_{invaLC} in Railway HPQC can then be deduced as in (2.40).

$$\left|\vec{V}_{invaLC}\right| = \sqrt{\left(V_{invaLCp}\right)^2 + \left(V_{invaLCq}\right)^2} = \sqrt{\left(V_{ac} - I_{caq} \cdot |X_{LCa}|\right)^2 + \left(I_{cap} \cdot |X_{LCa}|\right)^2} \tag{2.40}$$

It can then be observed from (2.40) that there can be conditions under which the value of V_{invaLC} can be lower than that of V_{ac}. An opposite voltage with V_{ac} is formed as additional voltage support when inductive reactive compensation current passes through the capacitive-coupled impedance in Railway HPQC. This allows the operation voltage in Railway HPQC to be less than the operation voltage in conventional RPC.

2.7.1.2 Comparisons Between RPC and HPQC Operation Voltage

Based on the analysis above, comparisons may be made between RPC and Railway HPQC.

Shown in Fig. 2.16 is a phasor diagram showing the operation difference between RPC and Railway HPQC. Under the same condition when the V_{ac} and V_{bc} phase compensation current, I_{ca} and I_{cb}, is output, it can be seen from the diagram that the operation voltage of Railway HPQC is lower than that of RPC.

By further investigating the diagram, it can be observed clearly that the direction of V_{La} in RPC is 90° anticlockwise to that of compensation current I_{ca}; while the direction of V_{LCa} in Railway HPQC is 90° clockwise to that of compensation current I_{ca}. Therefore, it causes additional voltage drop (V_{La} and V_{ac} are in the same direction) in RPC and voltage support (V_{LCa} and V_{ac} are in the opposite direction) in Railway HPQC. This may also be used to explain the reduction in Railway HPQC operation voltage.

As analyzed, the operation voltage requirement of RPC and Railway HPQC are given in (2.35) and (2.40), respectively. In order to simplify the analysis, some further manipulations are made.

The coupled impedance X_{La} and X_{LCa} is expressed as a ratio of the impedance when a voltage drop V_{ac} is resulted when compensation current I_{ca} passes through them. The relationship is expressed mathematically in (2.41) and (2.42), respectively. The value of the ratio is defined as m_{La} and m_{LCa}, respectively, in RPC and Railway HPQC.

$$X_{La} = m_{La} \cdot \frac{V_{ac}}{I_{ca}} \tag{2.41}$$

$$X_{LCa} = m_{LCa} \cdot \frac{V_{ac}}{I_{ca}} \tag{2.42}$$

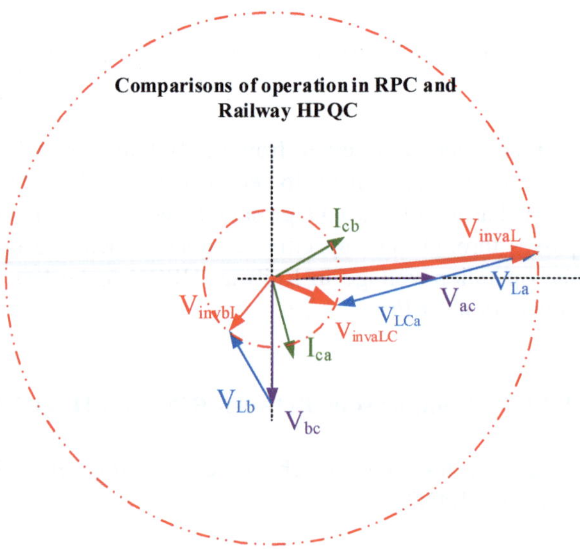

Fig. 2.16 Phasor diagram showing the comparisons between compensation operations in RPC and Railway HPQC

The active and reactive component of the compensation current, I_{cap} and I_{caq}, is further expressed as in (2.43) and (2.44).

$$I_{\mathrm{cap}} = I_{\mathrm{ca}} \cos \theta_{\mathrm{ca}} \tag{2.43}$$

$$I_{\mathrm{caq}} = I_{\mathrm{ca}} \sin \theta_{\mathrm{ca}} \tag{2.44}$$

By substituting (2.41), (2.42), (2.43), and (2.44) into (2.35) and (2.40), the expressions in (2.45) and (2.46) can be obtained. Notice that the effect of PCC voltage V_{ac} is eliminated by dividing the whole expressions by V_{ac}. The result obtained is regarded as the operation voltage rating k_{inv}, which is the ratio of operation voltage V_{inv} to PCC voltage V_{ac}, as shown in (2.45) and (2.46).

$$k_{\mathrm{invaL}} = \frac{\left|\vec{V}_{\mathrm{invaL}}\right|}{V_{\mathrm{ac}}} = \sqrt{(1 + \sin \theta_{\mathrm{ca}} \cdot m_{\mathrm{La}})^2 + (\cos \theta_{\mathrm{ca}} \cdot m_{\mathrm{La}})^2} \tag{2.45}$$

$$k_{\mathrm{invaLC}} = \frac{\left|\vec{V}_{\mathrm{invaLC}}\right|}{V_{\mathrm{ac}}} = \sqrt{(1 - \sin \theta_{\mathrm{ca}} \cdot m_{\mathrm{LCa}})^2 + (\cos \theta_{\mathrm{ca}} \cdot m_{\mathrm{LCa}})^2} \tag{2.46}$$

It can therefore be concluded that the operation voltage rating is dependent on the compensation angle (θ_{ca}) and the coupled impedance ratio (m_{La} or m_{LCa}). According to (2.17) and Table 2.4, under ideal compensation requirement (unbalance = 0%, PFA = PFB = PFC = 1.0), and the value of θ_{ca} is determined by (2.47). Thus, it can be further concluded that the operation voltage of power quality compensator (either RPC or Railway HPQC) is dependent on the load power factor (PF_L) and the coupled impedance ratio (m_{La} or m_{LCa}).

$$\theta_{\mathrm{ca}} = \tan\left(\frac{0.2887 \cdot PF_L + 0.5 \cdot \sqrt{(PF_L)^2 - 1}}{0.5 \cdot PF_L}\right) \tag{2.47}$$

For example, traction locomotive load power factor usually ranges from 0.8 to 0.9. A Matlab plot showing the operation voltage rating of RPC and Railway HPQC is constructed in Fig. 2.17 according to (2.45) and (2.46) and load power factor $PF_L = 0.85$. It can be clearly observed from the figure that the Railway HPQC operation voltage rating (k_{invaLC}) is always smaller than the conventional RPC operation voltage rating (k_{invaL}). This shows one great advantage of system with Railway HPQC.

2.7.2 *Further Analysis on Reduction Criteria*

Although it can be concluded from the analysis above that the usage of capacitive series LC branch coupling structure in Railway HPQC of the system can help to

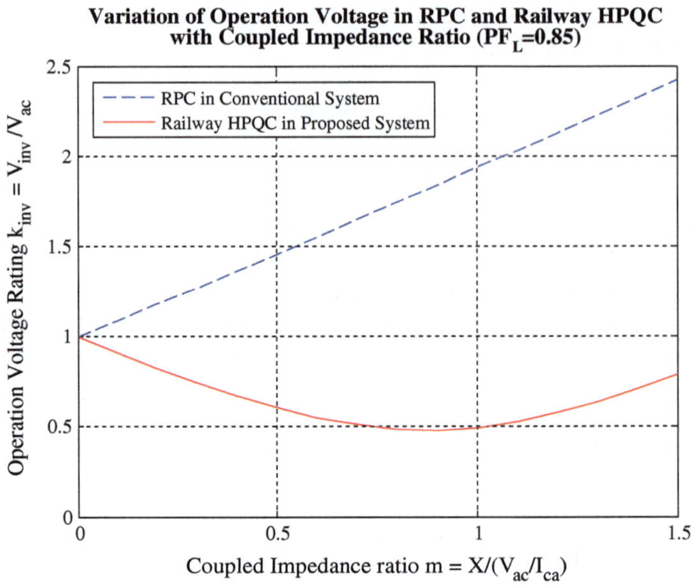

Fig. 2.17 A Matlab plot showing the variation of operation voltage in conventional RPC and Railway HPQC with coupled impedance ratio ($PF_L = 0.85$)

reduce the operation voltage during compensation, there are still certain criteria that not any capacitance value can be used. Some preliminary analyses are shown below.

Shown in Fig. 2.18 is a Matlab plot showing the variation of operation voltage rating in Railway HPQC with coupled impedance ratio m_{LCa} under three load power factor conditions ($PF_L = 0.8$, $PF_L = 0.85$, $PF_L = 0.9$). It can be observed that the curve of the Railway HPQC operation voltage rating is an inverted parabola, such that there will be conditions which value of k_{invaLC} will exceed 1. In other words, the operation voltage rating may exceed PCC voltage V_{ac}, like the case in conventional RPC.

This criteria limitation may be investigated mathematically, as shown in (2.48).

$$k_{\mathrm{invaLC}} = \sqrt{(1 - \sin\theta_{\mathrm{ca}} \cdot m_{\mathrm{LCa}})^2 + (\cos\theta_{\mathrm{ca}} \cdot m_{\mathrm{LCa}})^2} \leq 1 \tag{2.48}$$

The solution in (2.49) can then be obtained by solving (2.48). This is also the criterion for operation voltage reduction in the Railway HPQC.

$$0 \leq m_{\mathrm{LCa}} \leq 2\sin\theta_{\mathrm{ca}} \tag{2.49}$$

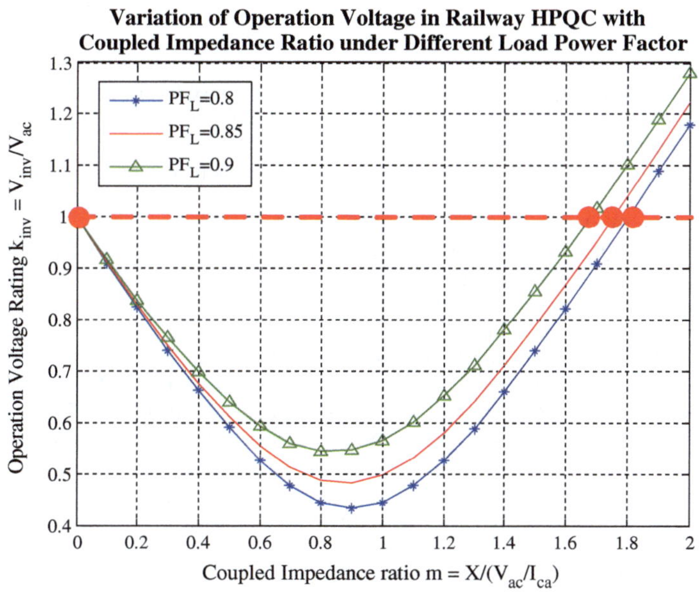

Fig. 2.18 A Matlab plot showing the boundary values of coupled impedance ratio in Railway HPQC which operation voltage rating is less than 1

Example

For example, with $PF_L = 0.80$, $PF_L = 0.85$ and $PF_L = 0.90$, the value of coupled impedance ratio m_{LCa} should not be larger than 1.8, 1.75, and 1.70, respectively, according to the computations based on (2.47) and (2.49).

2.8 Summary

In this chapter, the circuit structure of co-phase traction power supply system with Railway HPQC is introduced. The circuit parameters and power quality problems of co-phase traction power supply system are modeled so as to derive the control algorithm. The control is mainly divided into two parts: fundamental and harmonic compensation. The entire control algorithm is also developed and presented. In short, the required compensation power is calculated using instantaneous PQ theory and further manipulations according to power quality requirement. The reference current is then obtained using inverse Clarke transform and PWM signals are generated using hysteresis method. Hysteresis technique is selected since it is straightforward. Other PWM techniques may also be used. Comparisons are also made with co-phase traction with conventional RPC to show the advantages of system with Railway HPQC in operation voltage reduction.

References

1. Z. Shu, L. Xie and Q. Li, “Single-phase back-to-back converter for active power balancing, reactive power compensation and harmonic filtering in traction power system”, IEEE Trans. Power Electronics, Issue 99, 2010.
2. Zeliang Shu; Shaofeng Xie; and Qun-zhan Li, “Development and Implementation of a Prototype for Co-phase Traction Power Supply System”, *Asia-Pacific Power and Energy Engineering Conference (APPEEC)*, 2010, pp: 1–4.
3. Minwu Chen; Qunzhan Li; and Guang Wei, “Optimized Design and Performance Evaluation of New Cophase Traction Power Supply System”, *Asia-Pacific Power and Energy Engineering Conference (APPEEC)*, 2009, pp: 1–6.
4. Keng-Weng Lao; Man-Chung Wong; Ningyi Dai; Chi-Seng Lam; Lei Wang; Chi-Kong Wong, “Analysis of the Effects of Operation Voltage Range in Flexible DC Control on Railway HPQC Compensation Capability in High-Speed Co-phase Railway Power,” IEEE Transactions on Power Electronics, 2018, vol: 33, issue: 2, pp: 1760–1774.
5. Keng-Weng Lao; Man-Chung Wong; NingYi Dai; Chi-Seng Lam; Chi-Kong Wong; Lei Wang, “Analysis in the Effect of Co-phase Traction Railway HPQC Coupled Impedance on Its Compensation Capability and Impedance-Mapping Design Technique Based on Required Compensation Capability for Reduction in Operation Voltage”, 2017, vol: 32, issue: 4, pp: 2631–2646.
6. Keng-Weng Lao; Man-Chung Wong; Ning Yi Dai; Chi-Kong Wong; Chi-Seng Lam, “Analysis of DC-Link Operation Voltage of a Hybrid Railway Power Quality Conditioner and Its PQ Compensation Capability in High-Speed Cophase Traction Power Supply”, IEEE Transactions on Power Electronics, 2016, vol: 31, issue: 2, pp: 1643–1656.
7. Keng-Weng Lao; Man-Chung Wong; NingYi Dai; Chi-Kong Wong; Chi-Seng Lam, “A Systematic Approach to Railway Hybrid Quality Conditioner Design With Harmonic Compensation for High-Speed Railway”, *IEEE Transactions on Industrial Electronics*, 2015, vol: 62, issue: 2, pp: 930–942.
8. Ning Yi Dai; Keng-Weng Lao; Chi-Seng Lam, “Railway Hybrid Quality Conditioner With Partial Compensation for Converter Rating Reduction”, *IEEE Transactions on Industry Applications*, 2015, vol: 51, issue: 5, pp: 4130–4138.
9. Ning-Yi Dai; Man-Chung Wong; Keng-Weng Lao; Chi-Kong Wong, “Modelling and control of a railway power conditioner in co-phase traction power system under partial compensation”, *Power Electronics, IET*, 2014, vol: 7, issue: 5, pp: 1044–1054.
10. Lao, K.-W.; Dai, N.; Liu, W.-G.; Wong, M.-C., “Hybrid Power Quality Compensator With Minimum DC Operation Voltage Design for High-Speed Traction Power Systems”, *IEEE Transactions on Power Electronics*, 2013, vol: 28, issue: 4, pp: 2024–2036.
11. Keng-Weng Lao; NingYi Dai; WeiGang Liu; Man-Chung Wong; Chi-Kong Wong, “Modeling and control of Railway Static Power Conditioner compensation based on power quality standards”, *2012 IEEE 13th Workshop on Control and Modeling for Power Electronics (COMPEL)*, 2012, pp: 1–6.
12. Dai, N.Y.; Lao, K.W.; Wong, M.C.; Wong, C.K., “Hybrid power quality conditioner for co-phase power supply system in electrified railway”, *Power Electronics, IET*, 2012, vol: 5, issue: 7, pp: 1084–1094.

Chapter 3
Minimum Operation Voltage Design of Co-phase Traction Power with Railway HPQC for Steady Rated Load

3.1 Introduction

Different from Railway Power Quality Compensator (RPC) in [1–3], the Railway Power Quality Conditioner (Railway HPQC) possesses a minimum operation voltage point, which changes with the design of coupled impedance [4, 5]. Besides that, the ratio between inductance and capacitance in the Railway HPQC coupled impedance can contribute to further reduction in the operation voltage. In this chapter, the relationship between the coupled impedance in Railway HPQC and its operation voltage is investigated and presented in detail. The chapter starts with the investigations of the relationship between coupled impedance and operation voltage in Railway HPQC, followed by the discussions on the design of Railway HPQC for fundamental and harmonic compensation concern, respectively. The simulation and experimental results presented in this chapter support the validity of the design.

3.2 Relationship Between Co-phase Traction Power Quality Operation Voltage Rating and Other Parameters

In order to develop the system design of Railway HPQC for minimum inverter capacity rating, it is first important to understand the relationship between the co-phase power quality compensator operation voltage rating and other parameters. The power quality compensator operation voltage in this report refers to the dc link voltage of the back-to-back converter, which provides important voltage support for power transfer control during compensation. Since the dc link voltage is directly proportional to the inverter voltage V_{inv}, V_{inv} is mainly analyzed but will be calculated back as dc link voltage in final design procedure.

K.-W. Lao et al., *Co-phase Traction Power Supply with Railway Hybrid Power Quality Conditioner*, https://doi.org/10.1007/978-981-13-0438-5_3

3.2.1 Conventional System with RPC

Refer to the analysis in Chap. 2, the operation voltage rating requirement of RPC in conventional system is presented in (2.45). The RPC operation voltage for fundamental compensation is repeated for reference again in (3.1).

$$k_{\mathrm{invaL}} = \sqrt{(1 + \sin\theta_{\mathrm{ca}} \cdot m_{\mathrm{La}})^2 + (\cos\theta_{\mathrm{ca}} \cdot m_{\mathrm{La}})^2} \tag{3.1}$$

A 3D graph is plotted using Matlab according to (3.1) to show the variation of RPC operation voltage requirement with RPC coupled impedance m_{La} and load power factor PF_L. The following can be observed from the 3D plot:

- The load power factor has less influence on the RPC operation voltage rating requirement;
- The required RPC operation voltage rating increases almost linearly with RPC coupled impedance ratio m_{La};
- The minimum operation voltage rating occurs in conventional RPC only when the coupled impedance ratio m_{La} is zero, which indicates zero inductance, and is not possible in real condition (coupling inductor is always required).

3.2.2 Novel System with Railway HPQC

The required operation voltage rating in the system with Railway HPQC can be derived using similar method.

The Railway HPQC operation voltage rating is shown in (3.2).

$$k_{\mathrm{invaLC}} = \sqrt{(1 - \sin\theta_{\mathrm{ca}} \cdot m_{\mathrm{LCa}})^2 + (\cos\theta_{\mathrm{ca}} \cdot m_{\mathrm{LCa}})^2} \tag{3.2}$$

The 3D plot showing the variation of Railway HPQC operation voltage rating with Railway HPQC coupled impedance m_{La} and load power factor PF_L is presented in Fig. 3.2.

The following can be observed from the 3D plot in Fig. 3.2:

- The required Railway HPQC operation voltage rating increases for a larger load power factor PF_L;
- The required Railway HPQC operation voltage rating first increases with increasing coupled impedance ratio m_{LCa}, but then starts to decrease when the value of m_{LCa} exceeds a certain value;
- There exists a condition point at which Railway HPQC operation voltage rating is minimized when the value of coupled impedance ratio m_{LCa} is carefully selected.

Finally, by further comparing the graphics in Figs. 3.1 and 3.2, comparisons can be made between conventional RPC and novel Railway HPQC inverter capacity rating requirement:

- The novel Railway HPQC usually has lower operation voltage requirement than conventional RPC; this can be seen by the fact that all values in Fig. 3.1 are higher than 1.0, while the values in Fig. 3.2 are usually lower than 1.0;
- Different from conventional RPC, in the Railway HPQC, there is an operation point at which operation voltage rating is minimized and is achievable; this can be achieved so by carefully selecting the coupled impedance, which will be explained in the next section.

3.3 Minimum Operation Voltage Rating Design (Fundamental Compensation)

As analyzed in the previous section, in the system with Railway HPQC, minimum operation voltage rating requirement may be achieved by carefully designing the value of coupled impedance (represented by the ratio m_{LCa}). In this section, the

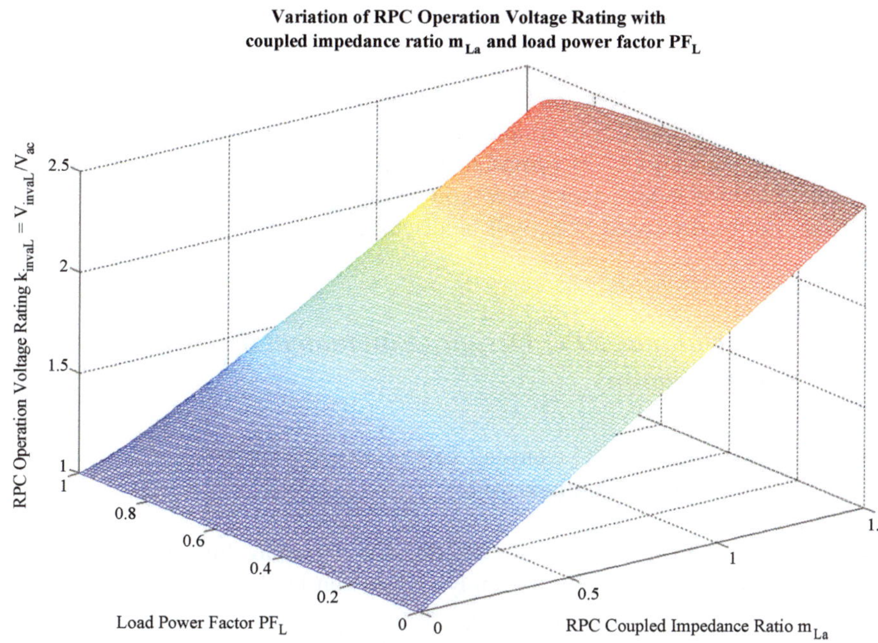

Fig. 3.1 A three-dimensional Matlab plot showing the variation of RPC operation voltage rating requirement with RPC coupled impedance m_{La} and load power factor PF_L

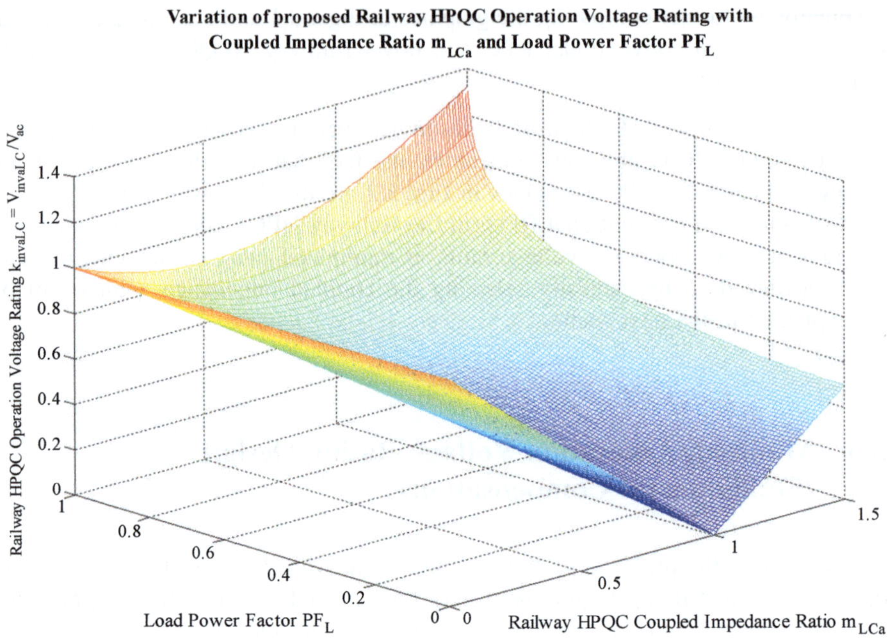

Fig. 3.2 A three-dimensional Matlab plot showing the variation of Railway HPQC operation voltage rating requirement with Railway HPQC coupled impedance m_{LCa} and load power factor PF_L

design of conventional RPC and novel Railway HPQC for minimum operation voltage rating is explored. For simpler analysis and easier understanding, it is first assumed that the majority of operation voltage rating is caused by fundamental (system unbalance and reactive power) compensation, harmonic compensation will, therefore, not be discussed here (will be discussed in next section).

3.3.1 Conventional RPC Design (Fundamental Compensation)

3.3.1.1 Conventional RPC Coupled Impedance Design (Fundamental Compensation)

The operation voltage of conventional RPC for providing fundamental compensation in co-phase traction power supply system can be determined by (2.35). By substituting the definition of RPC coupled impedance ratio m_{La} in (2.41) and the definitions of I_{cap} and I_{caq} in (2.43), (2.44) into the expression, (3.3) can be obtained

$$\left|\vec{V}_{\text{invaL}}\right| = \sqrt{\left(V_{\text{ac}} + I_{\text{ca\_rated}} \cdot \sin\theta_{\text{ca}} \cdot m_{\text{La}} \cdot \frac{V_{\text{ac}}}{I_{\text{ca\_rated}}}\right)^2 + \left(I_{\text{ca\_rated}} \cdot \cos\theta_{\text{ca}} \cdot \frac{V_{\text{ac}}}{I_{\text{ca\_rated}}}\right)^2} \tag{3.3}$$

The RPC fundamental compensation operation voltagerating can then be obtained by dividing the expression in (3.3) by V_{ac}, as shown in (3.4).

$$k_{\text{invaL}} = \frac{\left|\vec{V}_{\text{invaL}}\right|}{V_{\text{ac}}} = \sqrt{(1 + m_{\text{La}}\sin\theta_{\text{ca}})^2 + (m_{\text{La}}\cos\theta_{\text{ca}})^2} \tag{3.4}$$

According to (3.4), it could be observed that the RPC operation voltage rating for fundamental compensation is mainly dependent on the parameters compensation angle θ_{ca} and coupled inductance ratio m_{La}. The value of θ_{ca} is dependent on locomotive load power factor PF_L, as shown in (2.47); while the value of m_{La} is dependent on the coupled inductance L_a, which is usually determined based on the required low pass filter performance.

Here, more about the derivations of m_{La} or L_a are discussed. One major function of coupled inductance in inductive-coupled power quality compensator like RPC is to filter out the switching noises and harmonics introduced by the PWM controlled converter. The expression for such inductance (L_a) value determination is shown in. Detailed derivations are shown in [6] and will therefore not be derived in details here. The value of ΔI_{ca} is the output current ripple rms of the PWM controlled V_{ac} phase converter in conventional RPC.

$$L_a \geq \frac{V_{\text{dc}}}{8 f_s \Delta I_{\text{ca}}} \tag{3.5}$$

By substituting the value of V_{dc} by square root two times of the expression in (3.4) into (3.5), the expression in (3.6) can be obtained

$$L_a \geq \frac{\sqrt{2} \cdot \sqrt{(1 + m_{\text{La}}\sin\theta_{\text{ca}})^2 + (m_{\text{La}}\cos\theta_{\text{ca}})^2}}{8 f_s \Delta I_{\text{ca}}} \tag{3.6}$$

For smaller inductance and physical size, the minimum requirement of the inductance L_a is preferred. Thus, by further substitution of (2.41) into (3.6), (3.7) can be obtained.

$$L_a = \frac{m_{\text{La}} V_{\text{ac}}}{\omega I_{\text{ca\_rated}}} = \frac{\sqrt{2} \cdot V_{\text{ac}} \cdot \sqrt{(1 + m_{\text{La}}\sin\theta_{\text{ca}})^2 + (m_{\text{La}}\cos\theta_{\text{ca}})^2}}{8 f_s \Delta I_{\text{ca}}} \tag{3.7}$$

By solving m_{La} in (3.7), the expression in (3.8) can be obtained.

$$m_{\mathrm{La}} = \frac{-b \pm \sqrt{b^2 - 4ac}}{2a} \tag{3.8}$$

where
$$a = 32 \cdot f_s^2 \cdot \left(\frac{\Delta I_{\mathrm{ca}}}{I_{\mathrm{ca\_rated}}}\right)^2 - \omega^2$$
$$b = -2 \cdot \omega^2 \cdot \sin\theta_{\mathrm{ca}}$$
$$c = -\omega^2$$

Normally, the PWM switching frequency (f_s) is 10 kHz, and the value of $\Delta I_{\mathrm{ca}}/I_{\mathrm{ca\_rated}}$ represents the ripple ratio, and is usually kept within 10%. The variations of m_{La} with different values of $\Delta I_{\mathrm{ca}}/I_{\mathrm{ca\_rated}}$ and locomotive load power factor (PF_L) (in correspondence with θ_{ca}) are shown as a Matlab plot in Fig. 3.3 according to (3.8). It could be observed from the figure that the value of m_{LCa} does not change significantly when locomotive load power factor changes. In addition, when the value of $\Delta I_{\mathrm{ca}}/I_{\mathrm{ca\_rated}}$ is lower, the ripple allowed is less and the value of m_{La} is larger, indicating a larger coupled inductance requirement.

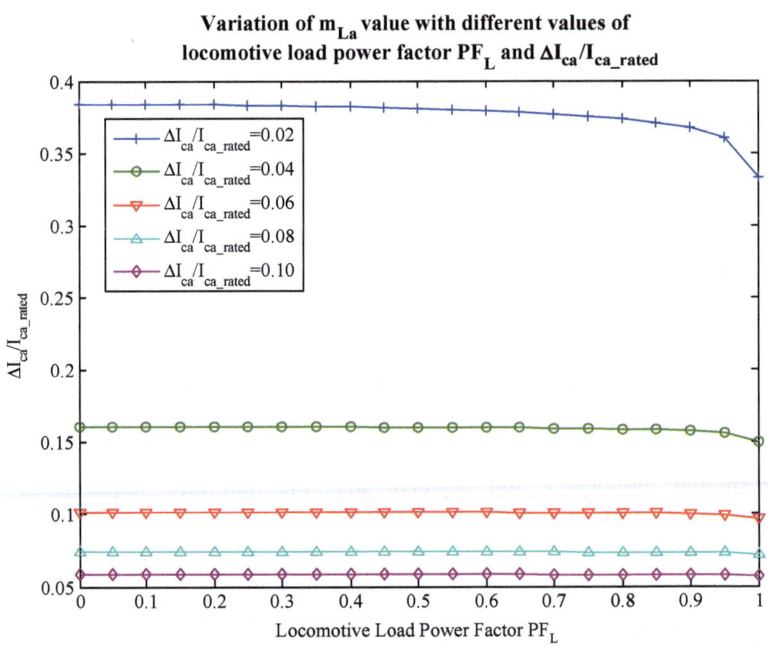

Fig. 3.3 Variation of the value of m_{La} with different values of locomotive load power factor PF_L and $\Delta I_{\mathrm{ca}}/I_{\mathrm{ca\_rated}}$

Conventional RPC Operation Voltage Rating (Fundamental Compensation)

By substituting the values of coupled inductance ratio m_{La} in Fig. 3.3 into (3.4), the conventional RPC operation voltage requirement for fundamental compensation under different locomotive load power factor PF_L and output current ripple limitation $\Delta I_{\mathrm{ca}}/I_{\mathrm{ca\_rated}}$ values can be determined. The graphic plot is shown in Fig. 3.4. Combining the observations from Figs. 3.3 and 3.4, the following can be concluded:

- When the ripple limitation is higher, the value of m_{La} and coupled inductance is also higher, and leads a high RPC operation voltage rating;
- With a higher locomotive power factor, the RPC operation voltage rating requirement is lower; however, the change is relatively small when locomotive load power factor changes within 0–0.9;
- For all the investigated conditions, the RPC operation voltage rating requirement is always higher than 1.0, and indicate that the RPC operation voltage should always be higher than the PCC voltage and induces high cost of high rating devices.

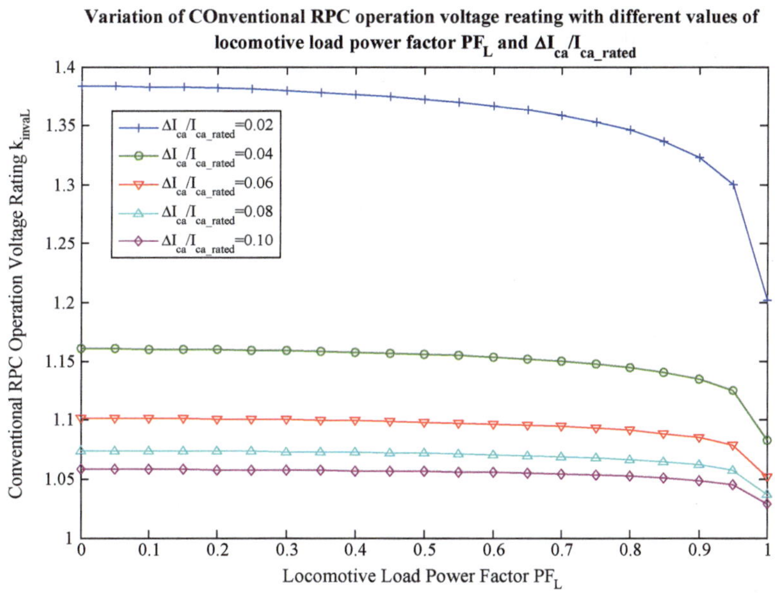

Fig. 3.4 A Matlab plot showing the variation of conventional RPC operation voltage rating for fundamental operation under different locomotive load power factor PF_L and output current ripple limitation $\Delta I_{\mathrm{ca}}/I_{\mathrm{ca\_rated}}$

3.3.2 *Novel Railway HPQC Design (Fundamental Compensation)*

Novel Railway HPQC Coupled Impedance Design (Fundamental Compensation)

Recall that fundamental compensation refers to system unbalance and reactive power compensation. The system parameter design is developed by further analyzing the Railway HPQC operation voltage rating requirement in (3.2).

A figure showing the variation of operation voltage requirement of Railway HPQC with different coupled impedance ratio m_{LCa} under load power factor value of 0.85 is shown in Fig. 3.5. As discussed and can be observed from the figure, in Railway HPQC, there is a point at which operation voltage rating is minimized. This point can be determined by solving the first derivative with respect to m_{LCa} of the expression in (3.2). With the assumption that the load condition is fixed and steady, the value of θ_{ca} is constant, and the result is shown in (3.9).

$$\frac{d(k_{\mathrm{invaLC}})}{d(m_{\mathrm{LCa}})} = 0 \Rightarrow -2\sin\theta_{\mathrm{ca}} + 2m_{\mathrm{LCa}} = 0 \tag{3.9}$$

By further manipulations, the value of m_{LCa} for minimum Railway HPQC operation voltage rating can be obtained, as shown in (3.10).

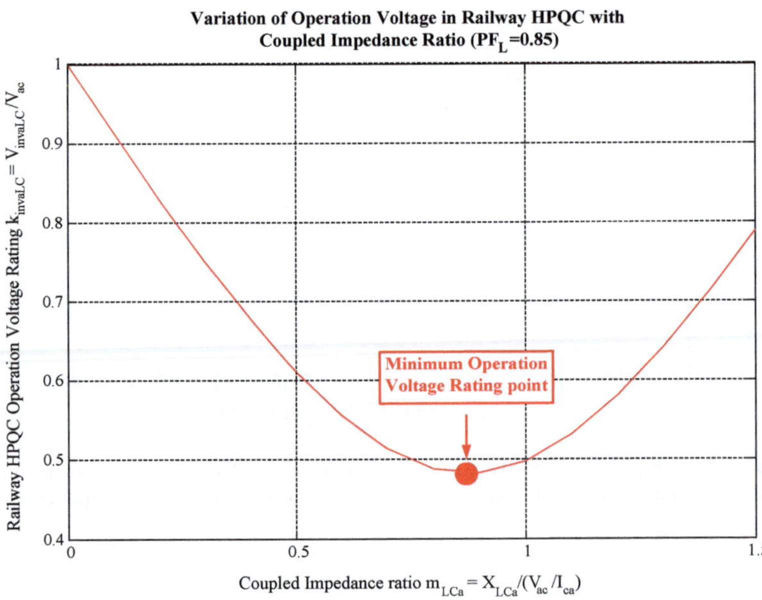

Fig. 3.5 A figure showing that there is a minimum operation voltage rating point in the Railway HPQC when the coupled impedance ratio value is carefully designed

$$\text{for minimum Railway HPQC opearation voltage rating,} \quad m_{\text{LCa}} = \sin\theta_{\text{ca}} \tag{3.10}$$

It can be seen from (3.10) that the coupled impedance is dependent on the load power factor PF_L. By substituting (2.42) into (3.10), the Railway HPQC coupled impedance X_{LCa} for minimum operation voltage rating can be determined by (3.11).

$$\text{for minimum Railway HPQC operation voltage rating,} \quad |X_{\text{LCa}}| = m_{\text{LCa}}\left(\frac{V_{\text{ac}}}{I_{\text{ca\_rated}}}\right) = \frac{V_{\text{ac}} \cdot \sin\theta_{\text{ca}}}{I_{\text{ca\_rated}}} \tag{3.11}$$

It can also be further investigated that the reactive power when the compensation current I_{ca} passes through the capacitive coupled impedance X_{LCa} is determined by (3.12).

$$Q_{\text{LCa}} = I_{\text{ca\_rated}} \cdot |X_{\text{LCa}}| = I_{\text{ca\_rated}} \cdot \frac{V_{\text{ac}} \cdot \sin\theta_{\text{ca}}}{I_{\text{ca\_rated}}} = V_{\text{ac}} \cdot \sin\theta_{\text{ca}} \tag{3.12}$$

It can be observed that under minimum Railway HPQC operation voltage rating the reactive compensation power is all provided by the capacitive coupled impedance X_{LCa}, This, in another sense, explains the reduction in Railway HPQC operation voltage rating since the converter in Railway HPQC can be designed to output active power only.

Railway HPQC Operation Voltage Rating (Fundamental Compensation)

The Railway HPQC coupled impedance design for minimum value of fundamental compensation operation voltage is already derived in (3.10) and (3.11). By substituting this value into the expression for determination of Railway HPQC operation voltage rating in (3.2), the expression for minimum fundamental compensation operation voltage can be deduced as in

$$k_{\text{invaLC}} = \sqrt{\left(1 - \sin^2\theta_{\text{ca}}\right)^2 + \left(\cos\theta_{\text{ca}} \cdot \sin\theta_{\text{ca}}\right)^2} \tag{3.13}$$

By further manipulations, it can be concluded that the minimum operation voltage rating achievable can be determined by (3.14). It can be observed that the minimum operation voltage rating is only dependent on the compensation power angle θ_{ca}. In fact, the minimum operation voltage for fundamental compensation equals to the ratio of active compensation power to the apparent compensation power.

$$k_{\text{invaLC\_min}} = \cos\theta_{\text{ca}} \tag{3.14}$$

In Fig. 3.6, the minimum operation voltage rating achievable for various locomotive power factor values is plotted using Matlab.

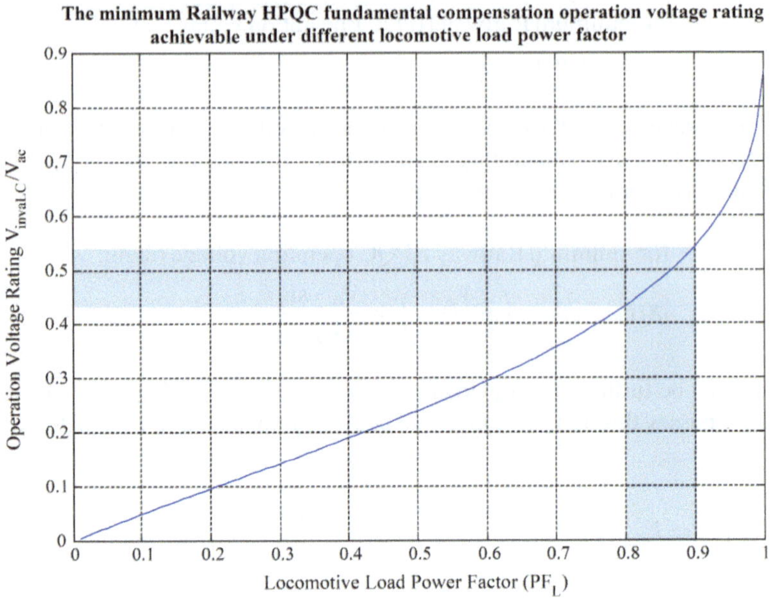

Fig. 3.6 A Matlab plot showing the minimum Railway HPQC operation voltage achievable under different locomotive load power factor PF_L

The following can be observed from the figure

- When the locomotive load power is higher, the minimum operation voltage for fundamental compensation using Railway HPQC is also higher;
- The relationship between minimum Railway HPQC fundamental compensation operation voltage and locomotive load power factor is not linear; as locomotive load power factor approaches unity (1.0), a small increase in load power factor leads to a very high increase in minimum operation voltage achievable.
- For all conditions, the Railway HPQC fundamental compensation operation voltage rating is always lower than 1.0, that is, lower than the conventional RPC ones.

Besides, as introduced previously, locomotive loadings usually ranges from 0.8 to 0.9. This area is shown shaded in Fig. 3.6. It can, therefore, be observed that the minimum operation voltage rating achievable ranges around 0.45–0.55.

In conventional RPC, in order to provide satisfactory compensation performance, the operation voltage must be larger than the PCC voltage. In other words, the RPC operation voltage must be always larger than 1.0 in order to provide satisfactory compensation performance. Comparing this value with 0.45–0.55, it can be found the largest reduction in the operation voltage is around 35–45% using Railway HPQC compared to conventional RPC. More details about operation voltage comparison will be given in the next section.

3.3.3 Reduction in Operation Voltage Rating (Fundamental Compensation)

In the previous sections, the design of co-phase traction power supply with conventional RPC and novel Railway HPQC for minimum operation voltage under fundamental compensation is explored. In this section, the reduction in operation voltage using Railway HPQC design compared to conventional RPC is discussed.

The operation voltage for conventional RPC and novel Railway HPQC for minimum operation voltage is shown in (3.4) and (3.14), respectively. The two equations are plotted together in Fig. 3.7, which is a combination of Figs. 3.4 and 3.6. It could be clearly observed that for all locomotive load power factor condition, the operation voltage of Railway HPQC is always lower than that of conventional RPC.

Example

For example, assuming that $PF_L = 0.80$, $PF_L = 0.85$ and $PF_L = 0.90$, and the current ripple limitation $\Delta I_{ca}/I_{ca\_rated}$ requirement is around 0.08, the operation voltage of conventional RPC and novel Railway HPQC can be computed according to (3.4) and (3.14).

A summary showing the reduction of operation voltage using Railway HPQC compared to conventional RPC under the investigated condition is shown in

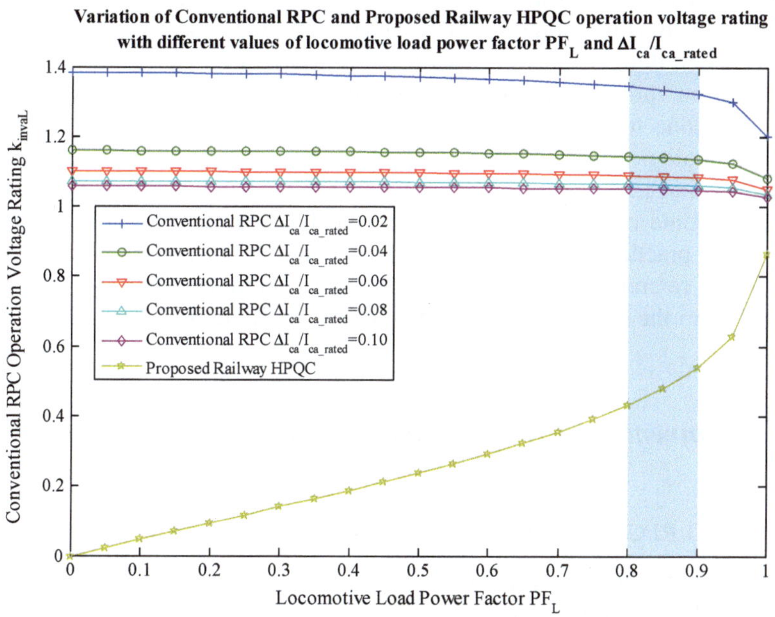

Fig. 3.7 Variation of conventional RPC and Railway HPQC operation voltage rating with different values of locomotive load power factor PF_L and $\Delta I_{ca}/I_{ca\_rated}$

Table 3.1 A summary comparing operation voltage rating of conventional RPC and novel Railway HPQC as well as reduction under-investigated load power factor conditions

	Conventional RPC		Novel railway HPQC		
PF_L	m_{La}	k_{invaL}	m_{LCa}	k_{invaLC}	Reduction (%)
0.80	0.0740	1.068	0.9010	0.5901	44.7
0.85	0.0739	1.066	0.8761	0.6071	43.0
0.90	0.0737	1.063	0.8397	0.6335	40.4

Table 3.1. It could be observed that for conventional RPC, the operation voltage decreases with increasing load power factor; on the other hand, for Railway HPQC, the operation voltage increases with load power factor. Under investigated conditions, there is at least 40% reduction in the operation voltage using novel Railway HPQC compared to conventional RPC.

3.4 Minimum Operation Voltage Rating Design (Harmonic Compensation)

So far, in the section above, the design is developed based on fundamental compensation (system unbalance and reactive); however, besides fundamental compensation, some additional portion of operation voltage needs to be added to provide harmonic compensations. In order to minimize the operation voltage rating, the harmonic compensation operation voltage should also be minimized. Similarly, analysis is first done on conventional RPC. In Railway HPQC, the analysis is a bit different from those done for fundamental compensations in previous section, analysis is done for two different designs, namely, the design based on existing HAPF technologies, and the design based on the method for minimum harmonic operation voltage.

For harmonic compensation, for easier analysis, investigations are performed based on the practical data obtained from the WuQing Traction Substation in China [1.53]. For reference, the on-site statistics of harmonic current distributions obtained from the WuQing substation is shown in Table 3.2.

3.4.1 Conventional RPC Design (Harmonic Compensation)

Conventional RPC Coupled Impedance Design (Harmonic Compensation)

As discussed previously, in conventional RPC, the coupled inductance is mainly determined by the current ripple limitation in its output compensation current, as shown in (3.5) to (3.8). There is, thus, less freedom in the design for minimizing operation voltage rating for harmonic compensation in conventional RPC.

Table 3.2 On-Site statistics of harmonic current distribution in WuQing substation traction load

	3rd	5th	7th	9th	11th
Harmonic contents (% of fundamental)	10.81	7.96	4.51	3.04	2.68

Conventional RPC Operation Voltage Rating (Harmonic Compensation)

In conventional RPC, a coupling inductor is used in the V_{ac} phase converter. With harmonic compensation defined as in (2.6) and (2.16), the operation voltage of conventional RPC for harmonic compensation is expressed as (3.15).

$$V_{\text{invaLh}} = \sqrt{\sum_{h=2}^{\infty} I_{\text{Lh\_h}}^2 X_{\text{Lah\_h}}^2} \tag{3.15}$$

The parameter $X_{\text{Lah\_h}}$ in (3.15) represents the impedance at hth harmonic frequency, as defined in (3.16).

$$X_{\text{Lah\_h}} = (h\omega)L_a \tag{3.16}$$

By substituting (2.41) (3.16) into (3.15), the expression in (3.17) is obtained.

$$V_{\text{invaLh}} = \sqrt{\sum_{h=2}^{\infty} I_{\text{Lh\_h}}^2 \left(h \cdot m_{\text{La}} \cdot \frac{V_{\text{ac}}}{I_{\text{ca\_rated}}} \right)^2} \tag{3.17}$$

With referring to the compensation core equation in (2.14), the expression in (3.17) can be further manipulated as (3.18). As introduced, for ideal power quality compensation, $K_1 = 0.5$, $K_2 = 0.2887$.

$$V_{\text{invaLh}} = \sqrt{\sum_{h=2}^{\infty} I_{\text{Lh\_h}}^2 \left(h \cdot m_{\text{La}} \cdot \frac{V_{\text{ac}}}{\left[\sqrt{(K_1 \cdot PF_L)^2 + \left(K_2 \cdot PF_L + \sqrt{1 - (PF_L)^2} \right)^2} \right] \cdot I_{\text{L\_rated}}} \right)^2} \tag{3.18}$$

A new parameter r_h is now introduced for further analysis. r_h refers to the ratio between harmonic component to fundamental one. For example, in Table 3.2, $r_3 = 0.1081$, $r_5 = 0.0796$, $r_7 = 0.0451$, $r_9 = 0.0304$, $r_{11} = 0.0268$. The expression in (3.18) can, therefore, be further simplified as (3.19). Similarly, the operation voltage is expressed as a ratio of PCC voltage V_{ac}, as operation voltage rating, k_{invaLh}.

$$k_{\text{invaLh}} = \frac{V_{\text{invaLh}}}{V_{\text{ac}}} = \sqrt{\sum_{h=2}^{\infty} (r_h)^2 \cdot \left(\frac{h \cdot m_{\text{La}}}{\sqrt{(K_1 \cdot PF_L)^2 + \left(K_2 \cdot PF_L + \sqrt{1 - (PF_L)^2} \right)^2}} \right)^2} \tag{3.19}$$

Referring to Table 3.2 and (3.19), the operation voltage for harmonic compensation under investigated condition is plotted in Fig. 3.8.

The following can be observed from the figure:

- Similar to conventional RPC fundamental compensation operation voltage rating, the conventional RPC harmonic compensation operation voltage also increases with increasing locomotive load power factor;
- Comparing to the graph of conventional RPC fundamental compensation operation voltage in Fig. 3.4, it could be observed that when the output current ripple limitation is more strict (like 2%), the harmonic operation voltage rating portion is not negligible.

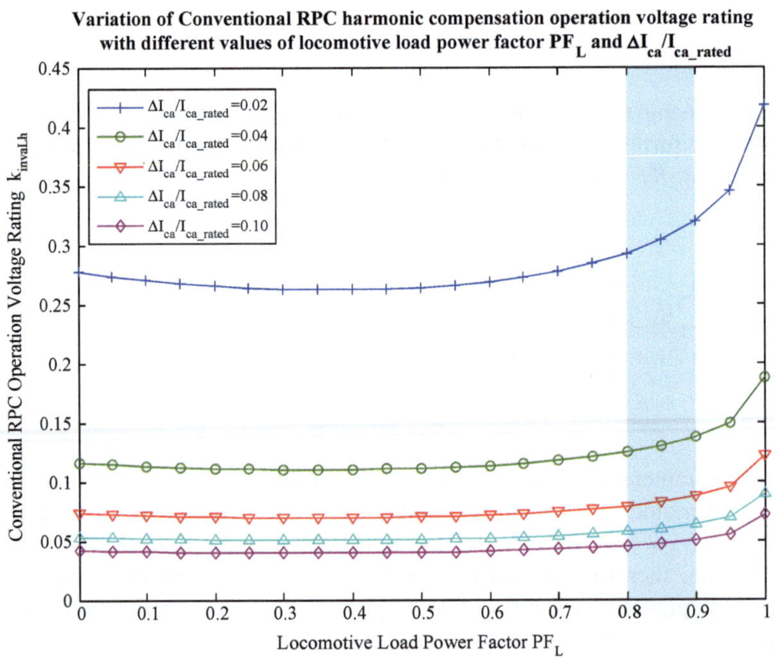

Fig. 3.8 Variation of conventional RPC harmonic compensation operation voltage rating with different values of locomotive load power factor PF_L and $\Delta I_{\text{ca}}/I_{\text{ca\_rated}}$

3.4.2 Novel Railway HPQC Using Traditional HAPF Design (Harmonic Compensation)

Next, the harmonic compensation operation voltage of novel Railway HPQC is being explored.

Railway HPQC Coupled Impedance Design using Traditional HAPF Method (Harmonic Compensation)

In contrast to conventional RPC, a hybrid series inductor–capacitor (LC) structure is used in the Railway HPQC, giving one more degree of freedom in designing coupled inductance and capacitance combination. First of all, the new railway HPQC using traditional HAPF design is investigated. The idea of hybrid LC structure design in Hybrid Active Power Filter (HAPF) for reactive power compensation only may be used. This method is being regarded as traditional HAPF design in the analysis that follows.

Traditionally, for HAPF, it is believed that when the resonance frequency of coupled inductance and capacitance is tuned at the harmonic frequency where most load harmonic contents are located at, the harmonic compensation operation voltage can be reduced and minimized. This mechanism works by that at resonance frequency, the overall impedance of coupled inductance and capacitance is zero and is thus short-circuit, it will, therefore, not cause any voltage drop.

The idea may also be understood in another way. According to the theory of Fourier Series, every periodic function can be decomposed into a sum of oscillating functions, such as sine and cosine functions. In short, a periodic function can be decomposed into different oscillating components of nth of fundamental frequency, in which n is an integer. As introduced in Chap. 1, harmonic contents refer to the frequency components other than the fundamental frequency one.

Referring to (2.6), the locomotive load harmonic contents can be expressed a sum of oscillating components; and the required output harmonic compensation current in V_{ac} phase converter equals load harmonic ones, the power quality compensator operation voltage for harmonic compensation, $V_{invaLCh}$, is, therefore, given by (3.20), in which X_{LCh_h} refers to the individual harmonic impedance of the capacitive coupled impedance at different harmonic frequencies, and will be discussed in details later.

$$V_{invaLCh} = \sqrt{\sum_{h=2}^{\infty} I_{Lh\_h}^2 X_{LCah\_h}^2} \tag{3.20}$$

Given a fixed steady loading condition, the values of I_{Ln} are fixed in (3.20). Therefore, investigations of harmonic compensation inverter capacity reduction is equivalent to reducing the individual harmonic impedance X_{LCah_h}.

In power system, harmonic current is usually introduced by nonlinear loadings such as nonlinear diode rectifier or power electronic components. The harmonic frequencies are, therefore, usually located at odd harmonics, 3rd, 5th, 7th, 9th, and so on, with most contents concentrated at lower harmonic frequencies. Traditional design for reducing power quality compensator operation voltage rating works in this way. Recall that the capacitive coupled structure is composed of a series LC branch. The overall coupled impedance of the series LC branch is kept fixed according to the reactive power compensation requirement, and the inductance and capacitance values are tuned to a certain harmonic frequency such that the series LC branch is made short-circuit at the tuned frequency to reduce the harmonic compensation workload, and thus the harmonic compensation operation voltage.

It may be a bit difficult to understand the idea by description only. For deeper understanding, the idea is analyzed mathematically. First of all, refer back to the system parameter definition of Railway HPQC in Fig. 2.13, the V_{ac} phase coupled inductance and capacitance are defined as L_a and C_{a}, respectively. The harmonic impedance at hth harmonic frequency, X_{LCah} is thus given by (3.21). Notice that the negative sign must be added before $X_{\mathrm{LCah\_h}}$ to represent the capacitive coupled characteristics of the LC branch.

$$-|X_{\mathrm{LCah\_h}}| = X_{\mathrm{Lah\_h}} + X_{\mathrm{Cah\_h}} = h\omega L_a - \frac{1}{h\omega \mathrm{C}_a} \tag{3.21}$$

For example, assuming that the series LC branch is tuned at nth harmonic frequency, the coupled capacitance and inductance value may be determined by setting nth harmonic impedance as zero, short-circuit, as shown in (3.22).

$$-|X_{\mathrm{LCah\_n}}| = n\omega L_a - \frac{1}{n\omega C_a} = 0 \tag{3.22}$$

However, it is not possible to solve for two unknowns, L_a and C_{a}, using one equation in (3.22) alone. Another important constraint equation would be that the overall fundamental coupled impedance should equal the designed X_{LCa}, such as the expression in (3.11). This constraint is shown mathematically in (3.23).

$$-|X_{\mathrm{LCa\_1}}| = \omega L_a - \frac{1}{\omega C_a} = -|X_{\mathrm{LCa}}| \tag{3.23}$$

Finally, the coupled inductance and capacitance may then be determined by solving (3.22) and (3.23) together, and the results are shown in (3.24).

$$\begin{cases} \mathrm{L}_a = \left(\frac{1}{\omega(n^2-1)}\right)|X_{\mathrm{LCa}}| \\ C_a = \frac{n^2-1}{n^2\omega|X_{\mathrm{LCa}}|} \end{cases} \tag{3.24}$$

By substituting the result in (3.24) into (3.21), the harmonic impedance of the series LC branch can be determined as (3.25).

$$X_{\mathrm{LCah\_h}} = \left(\frac{h^2 - n^2}{h(n^2 - 1)}\right)|X_{\mathrm{LCa}}| \tag{3.25}$$

For example, nonlinear loading harmonics are mostly concentrated at 3rd, 5th, and 7th harmonics, the harmonic impedance when the series LC branch is tuned at these frequencies are plotted in Fig. 3.9 using Matlab according to (3.25). For analysis, the values are expressed in ratio to its fundamental frequency impedance X_{LCa}. It can be observed from the figure that the series LC branch harmonic impedance is zero at the tuned frequency (n). Therefore, tuning of the series LC branch harmonic impedance at certain frequency can help to reduce the inverter capacity rating for harmonic compensation.

However, it can also be observed from (3.20) that the operation voltage for harmonic compensation does not only depend on the harmonic impedance, but also the load harmonic contents. Therefore, although tuning the series LC branch at a certain harmonic frequency can help to reduce harmonic compensation operation voltage, the operation voltage for harmonic compensation is not at a minimum value. Details about minimization of harmonic compensation operation voltage will be presented in the next subsection.

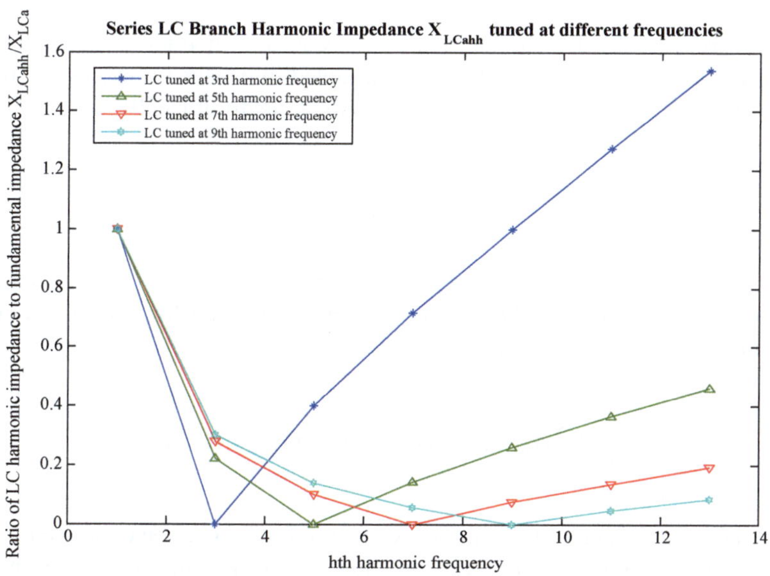

Fig. 3.9 A Matlab plot showing the ratio of power quality compensator series LC branch harmonic impedance to fundamental one when tuned at different frequencies

Railway HPQC Operation Voltage Rating using Traditional HAPF Design (Harmonic Compensation)

Similarly, the railway HPQC harmonic compensation operation voltage rating using traditional HAPF design is explored. In order to perform systematic analysis, further parameters k_L and k_C are defined.

According to (3.23), the expression for determination of the series LC branch fundamental impedance is revised as (3.42). For analysis, the fundamental impedance of coupling inductor $X_{\mathrm{La\_1}}$ and capacitor $X_{\mathrm{Ca\_1}}$ is expressed in k_L and k_C times of X_{LCa}. These two parameters are important in the design analysis.

$$-|X_{\mathrm{LCa}}| = X_{\mathrm{La}}\_1 + X_{\mathrm{Ca}}\_1 = (k_L + k_c)|X_{\mathrm{LCa}}| \tag{3.26}$$

The expression for individual harmonic impedance of the series LC branch is revised as (3.27) by substituting (3.42) into (3.25)

$$X_{\mathrm{LCah\_h}} = -\left(hk_L + \frac{1}{h}k_C\right)|X_{\mathrm{LCa}}| \tag{3.27}$$

By further substituting (3.43) into (3.27), the expression can be further revised as (3.28).

$$X_{\mathrm{LCah\_h}} = -\frac{1}{h}\left[(h^2 - 1)k_L - 1\right] \cdot |X_{\mathrm{LCa}}| \tag{3.28}$$

Recall from (3.20) that the harmonic compensation inverter capacity is dependent on the load harmonics and harmonic coupled impedance of series LC branch. For reference, this important expression is shown again in (3.29).

$$V_{\mathrm{invaLCh}} = \sqrt{\sum_{h=2}^{\infty} I^2_{\mathrm{Lh\_h}} X^2_{\mathrm{LCah\_h}}} \tag{3.29}$$

By substituting the harmonic distribution model in (3.41) and the harmonic impedance expression in (3.27) into (3.29), the expression for harmonic compensation inverter capacity is obtained as

$$V_{\mathrm{invaLCh}} = \sqrt{\sum_{h=2}^{\infty} (r_{\mathrm{h}}\_\mathrm{h}I_{L1})^2 \left[-\frac{1}{h}[(h^2 - 1)k_L - 1] \cdot |X_{\mathrm{LCa}}|\right]^2} \tag{3.30}$$

Based on (2.17) and Table 2.4, the expression for I_{L1} may be expressed in terms of rated compensation current $I_{\mathrm{ca\_rated}}$, as shown in (3.31). Since the denominator all refer to constant values for steady fixed load condition, therefore it is defined as A in the analysis that follows.

$$I_{L1} = \frac{I_{\mathrm{ca\_rated}}}{\sqrt{\left(K_2 PF_L + \sqrt{PF_L^2 - 1}\right)^2 + (K_1 PF_L)^2}} = \frac{I_{\mathrm{ca\_rated}}}{A} \tag{3.31}$$

where $K_1 = 0.5$, $K_2 = 0.2887$ for ideal power quality compensation

The expression of I_{L1} in (3.31) can be substituted into (3.30) to obtain the final expression for harmonic compensation operation voltage in (3.32).

$$V_{\mathrm{invaLCh}} = \sqrt{\sum_{h=2}^{\infty} \left(r_{\mathrm{h\_h}} \cdot \frac{I_{\mathrm{ca\_rated}}}{A}\right)^2 \left[-\frac{1}{h}[(h^2 - 1)k_L - 1] \cdot |X_{\mathrm{LCa}}|\right]^2} \tag{3.32}$$

As discussed in previous section, in Railway co-phase traction power supply, for minimum fundamental compensation operation voltage, the V_{ac} phase coupled impedance X_{LCa} is designed according to (3.11). By substituting (3.11) into (3.32), the expression can be further revised as (3.33).

$$V_{\mathrm{invaLCh}} = \sqrt{\sum_{h=2}^{\infty} \left(r_{\mathrm{h\_h}} \cdot \frac{I_{\mathrm{ca\_rated}}}{A}\right)^2 \left[-\frac{1}{h}[(h^2 - 1)k_L - 1] \cdot \frac{V_{\mathrm{ac}} \sin\theta_{\mathrm{ca}}}{I_{\mathrm{ca\_rated}}}\right]^2} \tag{3.33}$$

The operation voltage V_{invaLCh} in (3.34) may then be obtained by further manipulations. For analysis, similar to previous analysis, the inverter capacity rating k_{invaLCh} is defined as ratio of X_{invaLCh} to rated capacity $V_{\mathrm{ac}} I_{\mathrm{ca\_rated}}$, as shown in (3.35).

$$V_{\mathrm{invaLCh}} = \sqrt{\sum_{h=2}^{\infty} r_{\mathrm{h\_h}}^2 \left[\frac{(V_{\mathrm{ac}} \sin\theta_{\mathrm{ca}})^2}{A^2}\right] \left[-\frac{1}{h}[(h^2 - 1)k_L - 1]\right]^2} \tag{3.34}$$

$$k_{\mathrm{invaLCh}} = \frac{V_{\mathrm{invaLCh}}}{V_{\mathrm{ac}}} = \sqrt{\sum_{h=2}^{\infty} (r_{\mathrm{h\_h}}^2) \left[\frac{(\sin\theta_{\mathrm{ca}})^2}{A^2}\right] \left[-\frac{1}{h}[(h^2 - 1)k_L - 1]\right]^2} \tag{3.35}$$

As discussed, for LC branch design in Railway HPQC using traditional HAPF method, the LC branch is usually tuned at certain harmonic frequency (for example, at h harmonics), such that the relationship between coupled inductance L_a and coupled capacitance C_a satisfies the expression in (3.36)

$$L_a C_a = \frac{1}{h^2 \omega^2} \tag{3.36}$$

The definition of k_L and k_C can be found in (3.37).

$$-|X_{\mathrm{LCa}}| = X_{\mathrm{La\_1}} + X_{\mathrm{Ca\_1}} = (k_L + k_c)|X_{\mathrm{LCa}}| \tag{3.37}$$

It can be inferred from (3.37) that the value of k_L is determined by (3.38).

$$k_L = \frac{X_{\mathrm{La\_1}}}{|X_{\mathrm{LCa}}|} = \frac{\omega L_a}{-\left(\omega L_a - \frac{1}{\omega C_a}\right)} \tag{3.38}$$

By substituting (3.36) into (3.38), the value of k_L in Railway HPQC using traditional HAPF method (tuned at certain, h harmonic frequency) can be derived as in (3.39). In order to distinguish the tuned harmonic from the individual harmonics in the calculation, a parameter n is used to replace the value of h in the expression.

$$k_L = \frac{1}{h^2 - 1} = \frac{1}{n^2 - 1} \tag{3.39}$$

By substituting (3.39) into (3.35), the expression for determination of Railway HPQC harmonic compensation operation voltage rating using traditional HAPF LC branch design is shown in (3.40).

$$k_{\mathrm{invaLCh}} = \sqrt{\sum_{h=2}^{\infty} \left(r_{\mathrm{h\_h}}^2\right) \left[\frac{(\sin\theta_{\mathrm{ca}})^2}{A^2}\right] \left[-\frac{1}{h}\left[(h^2 - 1)\left(\frac{1}{n^2 - 1}\right) - 1\right]\right]^2} \tag{3.40}$$

A Matlab graph is then plotted in Fig. 3.10 based on (3.40) to show the variation of Railway HPQC operation voltage rating for harmonic compensation under different load power factor and tuned harmonic frequency.

The following can be observed from the figure:

- The curves in the figure are all inverted parabolas, showing the relationship between the Railway harmonic compensation operation voltage rating and locomotive load power factor PF_L is nonlinear and is nor mono-increasing or mono-decreasing function;
- The Railway HPQC operation voltage for harmonic compensation goes from highest to lowest: design to be tuned at 5th harmonics, tuned at 7th harmonics, tuned at 3rd harmonics; this is different from the general idea that when the LC filter is tuned at the harmonic frequency which load harmonics are mostly concentrated at, the harmonic compensation operation voltage will be relatively low.

This shows that Railway HPQC using conventional LC filter design may not be able to minimize the operation voltage. The design is better to be related to load harmonic distribution condition.

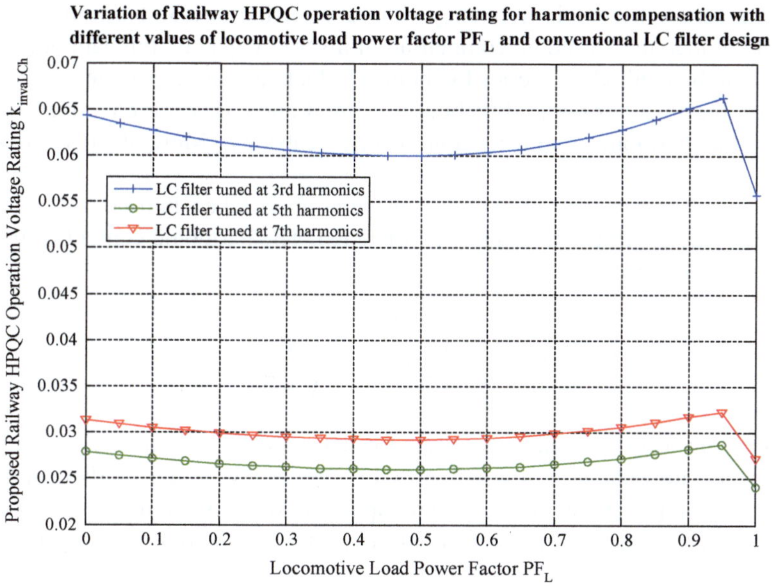

Fig. 3.10 A graph showing the variation of Railway HPQC harmonic compensation operation voltage rating under different locomotive load power factor PF_L and tuned harmonic frequency of the LC filter

For example, in the WuQing substation, the most dominant load harmonic contents are located at 3rd harmonics, using traditional HAPF design, the hybrid LC branch would be tuned at 3rd harmonic frequency. It could be observed from Fig. 3.10 that different from the idea of traditional HAPF LC branch design, the harmonic compensation operation voltage rating is not minimized when the LC branch is tuned at 3rd harmonic frequency (most dominant load harmonic contents). Therefore, in order to minimize the Railway HPQC harmonic operation voltage rating, the design of LC branch should not be dependent only on the dominant load harmonic frequency but also should consider the entire load harmonics distribution. More will be explored in the next section.

3.4.3 Railway HPQC Using New Design Method (Harmonic Compensation)

Railway HPQC Impedance Design using New Method (Harmonic Compensation)

As explained in previous contents, the Railway HPQC harmonic compensation is not only dependent on the harmonic impedance of the series LC branch but also the load harmonic distribution. Therefore, traditional method of tuning the series LC branch at certain harmonic frequency without referring to load harmonic condition can only reduce the harmonic compensation inverter capacity, but cannot minimize it. Thus, in the method, the series LC branch inductance and capacitance is designed based on the load harmonic distribution method and can work effectively in minimizing the harmonic compensation inverter capacity. More details are shown below.

First of all, since load harmonic consideration is taken into consideration, the load harmonic distribution needs to be modeled. The harmonic content distribution is represented as (3.41).

$$I_{\mathrm{cah}} = I_{\mathrm{Lh}} = \sqrt{\sum_{h=2}^{\infty} I_{\mathrm{Lh\_h}}^{2}} = \sqrt{\sum_{h=2}^{\infty} \left(r_{\mathrm{h\_h}} I_{L1}\right)^{2}} \tag{3.41}$$

According to (3.23), the expression for determination of the series LC branch fundamental impedance is revised as (3.42). For analysis, the fundamental impedance of coupling inductor $X_{\mathrm{La\_1}}$ and capacitor $X_{\mathrm{Ca\_1}}$ is expressed in k_L and k_C times of X_{LCa}. These two parameters are important in the design analysis.

$$-|X_{\mathrm{LCa}}| = X_{\mathrm{La\_1}} + X_{\mathrm{Ca\_1}} = (k_L + k_c)|X_{\mathrm{LCa}}| \tag{3.42}$$

By further investigating (3.42), the relationship between k_L and k_C can be derived, as shown in (3.43).

$$k_L + k_c = -1 \tag{3.43}$$

A Matlab plot showing different possible combinations of k_L and k_C values according to (3.43) is shown in Fig. 3.11. Since the coupled inductance must be inductive, the value of k_L cannot be negative; similarly, the coupled capacitance must be capacitive, therefore the value of k_C cannot be positive. These not-possible regions are shown as shaded area in the figure.

As can be seen from Fig. 3.11, there are still different possible combinations and the method of choosing the combination which yields minimum harmonic compensation operation voltage will be introduced below.

The Railway HPQC harmonic compensation operation voltage rating is shown in (3.35). The value of k_L in which the harmonic compensation operation voltage rating k_{invaLCh} is at its minimum may, thus, be derived by (3.44), which is obtained by setting the first derivative of k_{invaLCh} with respect to k_L equals zero.

$$\frac{dk_{\mathrm{invaLCh}}^{2}}{dk_L} = \sum_{h=2}^{\infty}\left(r_{\mathrm{h\_h}}^{2}\right)\left[\frac{(\sin\theta_{\mathrm{ca}})^{2}}{A^{2}}\right]\cdot 2\left[-\frac{1}{h^{2}}\left(h^{2}-1\right)^{2}k_L + \frac{1}{h^{2}}\left(h^{2}-1\right)\right] = 0 \tag{3.44}$$

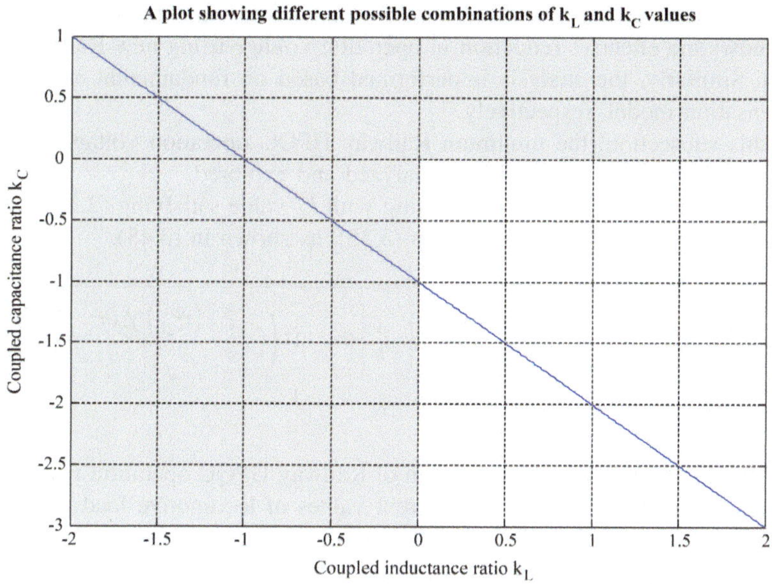

Fig. 3.11 A Matlab plot showing different possible combinations of k_L and k_C values according to (3.43)

The expression for determination of k_L value that can yield minimum operation voltage for harmonic compensation is obtained by further manipulation of (3.44), as shown in (2.26).

$$k_L = \frac{\sum_{h=2}^{\infty}\left(r_{\text{h\_h}}^2\right)\left[\frac{1}{h^2}\left(h^2-1\right)\right]}{\sum_{h=2}^{\infty}\left(r_{\text{h\_h}}^2\right)\left[\frac{1}{h^2}\left(h^2-1\right)^2\right]} \tag{3.45}$$

Knowing the value of coupled inductance ratio k_L, the coupled inductance may then be determined by the expression in (3.46).

$$L_a = \frac{|X_{\text{La}}|}{\omega} = \frac{k_L|X_{\text{LCa}}|}{\omega} \tag{3.46}$$

The coupled capacitance can then be determined by the expression in (3.47).

$$C_a = \frac{1}{\omega|X_{Ca}|} = \frac{1}{\omega \cdot k_c|X_{LCa}|} = \frac{1}{\omega \cdot (-1-k_L)|X_{LCa}|} \tag{3.47}$$

Railway HPQC Operation Voltage Rating using New Method (Harmonic Compensation)

The parameter design of Railway HPQC for minimum operation voltage is derived in previous two sections. In this subsection, the minimum operation voltage rating achievable using new design under steady fixed load condition is analyzed. This shows the effective reduction in operation voltage using new Railway HPQC design. Similarly, the analysis is performed based on fundamental and harmonic compensation model, respectively.

In this subsection, the minimum Railway HPQC operation voltage achievable for harmonic compensation will be analyzed and explored.

The harmonic operation voltage rating with k_L value satisfying (2.26) may then be obtained by substituting (3.45) into (3.35), as shown in (3.48).

$$k_{\text{invaLCh}} = \frac{V_{\text{invaLCh}}}{V_{\text{ac}}} = \sqrt{\sum_{h=2}^{\infty}\left(r_{\text{h\_h}}^2\right)\left[\frac{(\sin\theta_{\text{ca}})^2}{A^2}\right]\left[-\frac{1}{h}\left[(h^2-1)\left(\frac{\sum_{h=2}^{\infty}\left(r_{\text{h\_h}}^2\right)\left[\frac{1}{h^2}(h^2-1)\right]}{\sum_{h=2}^{\infty}\left(r_{\text{h\_h}}^2\right)\left[\frac{1}{h^2}(h^2-1)^2\right]}\right)-1\right]\right]^2} \tag{3.48}$$

A Matlab plot showing the variation of Railway HPQC operation voltage rating for harmonic compensation with different values of locomotive load power factor PF_L under the LC filter design. The following can be observed from the figure:

- Similar to conventional LC filter design, the curve is inverter parabola, meaning that the relationship between locomotive load power factor and harmonic compensation operation voltage is nonlinear;
- However, different from conventional LC filter design, with the new LC filter design, the Railway HPQC operation voltage rating for harmonic compensation can be very low (lower than 0.03, refer to graph obtained from conventional design in Fig. 3.10 showing all values higher than 0.5).

3.4.4 Reduction in Operation Voltage Rating (Harmonic Compensation)

In the following, the operation voltage rating for harmonic compensation in (a) conventional RPC; (b) new Railway HPQC using traditional HAPF method; and (c) new Railway HPQC using new method are being compared and analyzed.

As explored, the harmonic compensation operation voltage rating for the three conditions are shown in (3.19), (3.40), and (3.48), as well as Figs. 3.8, 3.10, and 3.12, respectively. These three graphic plots are combined together in Fig. 3.13 to do further comparisons. It is shown clearly in the figure that by using new LC branch design method in Railway HPQC, the harmonic compensation operation voltage rating can be effectively reduced.

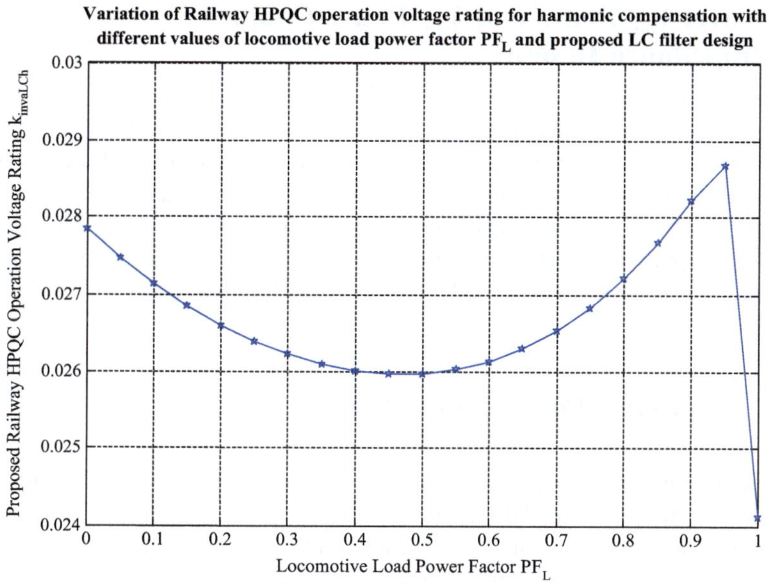

Fig. 3.12 A graph showing the variation of Railway HPQC operation voltage rating for harmonic compensation with different value of locomotive load power factor PF_L under new LC filter design

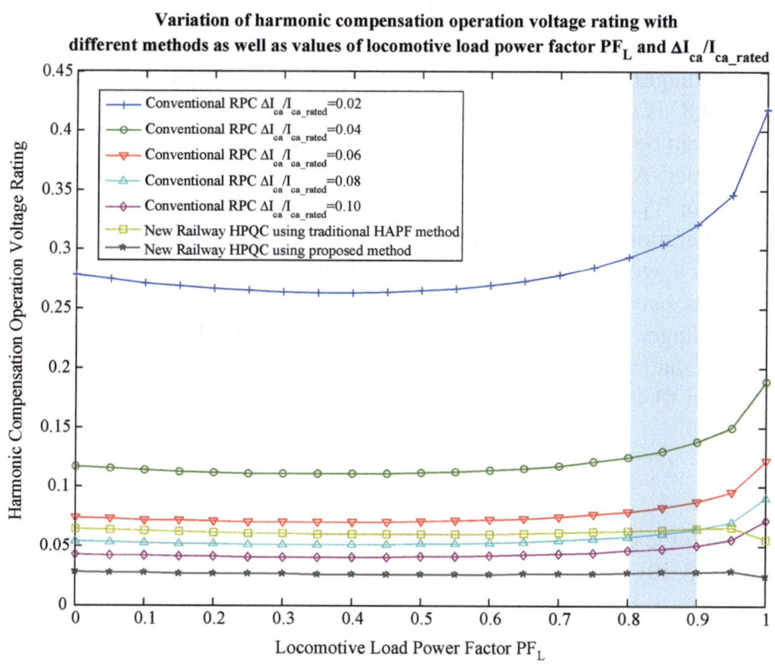

Fig. 3.13 Variation of harmonic compensation operation voltage rating with different methods to show the minimization effect using new LC filter method in Railway HPQC

3.5 Novel V_{ac} Phase and V_{bc} Phase Coupled Impedance Design

In the analysis above, it has been assumed that the operation voltage rating of the traction power quality compensator, no matter conventional RPC or Railway HPQC, is dominant by the V_{ac} phase converter. So far, only the coupled impedance design of V_{ac} phase converter is studied, and there is less discussion on the coupled impedance design of V_{bc} phase converter. In order to provide a clearer picture, the design of V_{ac} phase and V_{bc} phase coupled impedance are covered in this section.

3.5.1 V_{ac} *Phase Coupled Impedance*

Detailed definitions of V_{ac} phase converter parameters are shown in Fig. 2.13. According to the Railway HPQC design based on the requirement of minimum operation voltage rating, the V_{ac} phase coupled impedance X_{LCa} is selected according to (3.11). The coupled inductance L_a and capacitance C_a may then be determined by (3.45), (3.46), and (3.47).

3.5.2 V_{bc} *Phase Coupled Impedance*

The vector diagram showing the operation of the V_{bc} phase converter in the Railway HPQC is shown in Fig. 3.14. Notice that the value of V_{bc} is lower than that of V_{ac}. This can be easily achieved by reducing the turning ratio of the V_{bc} phase in V/V transformer. Assuming a fixed V_{bc} compensation current, and thus fixed vector I_{cb}, the vector V_{LCb} varies along the line L_1 with different values of X_{Lb}. The minimum operation voltage boundary of the V_{ac} phase converter is represented by the circle Cir_a, with radius of $V_{\mathrm{invaLC\_min}}$. It is obvious that there are two intersection points between Cir_a and L_1, which satisfy the value of minimum HPQC operation voltage. They may be determined mathematically.

The mathematical expression showing the intersections of circle Cir_a and the line L_2 is given in (3.49).

$$V^2_{\mathrm{invaLC\_min}} = \left(V^2_{\mathrm{LCb}} \sin^2\theta_{\mathrm{cb}} + (V_{\mathrm{bc}} - V_{\mathrm{LCb}}\cos\theta_{\mathrm{cb}})^2\right) \tag{3.49}$$

By solving the expression, the mathematical expressions for pt.1 and pt.2 can be obtained in (3.50).

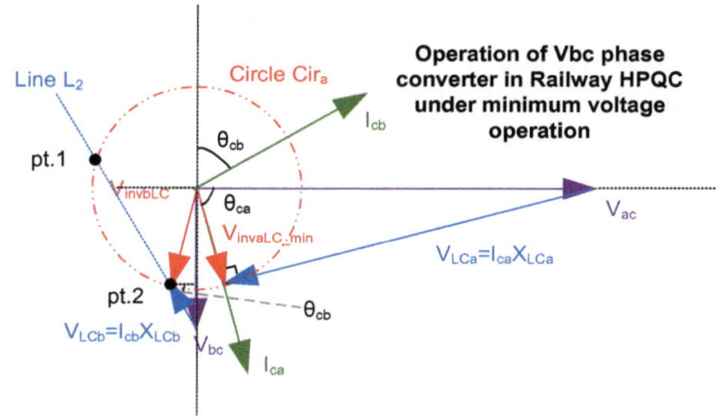

Fig. 3.14 The selection of V_{bc} coupled impedance for V_{ac} phase inverter rating matching

$$X_{\mathrm{LCb}} = \frac{V_{\mathrm{bc}} \sin \theta_{cb} + \sqrt{V_{\mathrm{invaLC\_min}}^2 - V_{\mathrm{bc}}^2 \cos^2 \theta_{\mathrm{cb}}}}{I_{cb}} \text{(pt.1)}$$
$$X_{\mathrm{LCb}} = \frac{V_{\mathrm{bc}} \sin \theta_{cb} - \sqrt{V_{\mathrm{invaLC\_min}}^2 - V_{\mathrm{bc}}^2 \cos^2 \theta_{\mathrm{cb}}}}{I_{\mathrm{cb}}} \text{(pt.2)} \quad (3.50)$$

Although pt.1 and pt.2 may both satisfy the voltage matching with $V_{\mathrm{invaLC\_min}}$, operation point at pt.2 is preferred due to the lower impedance of X_{LCb} and lower power consumptions.

3.6 Comprehensive Design Procedure for Minimum Railway HPQC Operation Voltage

Based on the discussions above, the comprehensive design procedure for minimum Railway HPQC operation voltage is given below.

1. Obtain and compute the value of locomotive PCC voltage V_{ac}, rated compensation current $I_{\mathrm{ca\_rated}}$, and V_{ac} phase compensation angle θ_{ca};
2. Determine the value of V_{ac} phase coupled impedance ratio m_{LCa} according to (3.11);
3. Select the value of V_{ac} phase coupled impedance X_{LCa} according to (3.51);

$$|X_{\mathrm{LCa}}| = m_{\mathrm{LCa}} \cdot \frac{V_{ac}}{I_{\mathrm{ca\_rated}}} = \frac{V_{\mathrm{ac}} \sin \theta_{\mathrm{ca}}}{I_{\mathrm{ca\_rated}}} \quad (3.51)$$

4. For harmonic compensation concern, the coupled inductance ratio k_L may be computed based on (3.45);
5. Determine the coupled inductance L_a and capacitance C_a according to the expression in (3.52) and (3.53);

$$L_a = \frac{k_L|X_{\mathrm{LCa}}|}{\omega} = \left(\frac{\sum_{h=2}^{\infty}\left(r_{\mathrm{h\_h}}^2\right)\left[\frac{1}{h^2}(h^2-1)\right]}{\sum_{h=2}^{\infty}\left(r_{\mathrm{h\_h}}^2\right)\left[\frac{1}{h^2}(h^2-1)^2\right]}\right)\cdot\left(\frac{|X_{\mathrm{LCa}}|}{\omega}\right) \tag{3.52}$$

$$C_a = \left(-1 - \frac{\sum_{h=2}^{\infty}\left(r_{\mathrm{h\_h}}^2\right)\left[\frac{1}{h^2}(h^2-1)\right]}{\sum_{h=2}^{\infty}\left(r_{\mathrm{h\_h}}^2\right)\left[\frac{1}{h^2}(h^2-1)^2\right]}\right)\cdot\left(\frac{1}{\omega\cdot|X_{\mathrm{LCa}}|}\right) \tag{3.53}$$

6. Determine the V_{bc} phase coupled impedance and inductance by (3.54) and (3.55);

$$|X_{\mathrm{LCb}}| \geq \frac{V_{\mathrm{bc}}\sin\theta_{\mathrm{cb}} - \sqrt{V_{\mathrm{invaLC\_min}}{}^2 - V_{\mathrm{bc}}^2\cos^2\theta_{\mathrm{cb}}}}{I_{\mathrm{cb}}} \tag{3.54}$$

$$L_b = \frac{|X_{LCb}|}{\omega} \tag{3.55}$$

7. Compute and determine the operation voltage rating of Railway HPQC, k_{invaLC}, according to (3.13) and (3.48);
8. Compute the Railway HPQC dc Link Voltage according to (3.56)

$$V_{\mathrm{dc}} = \sqrt{2\left\{(1-m_{\mathrm{LCa}}\sin\theta_{\mathrm{ca}})^2 + (m_{\mathrm{LCa}}\cos\theta_{\mathrm{ca}})^2 + \sum_{h=2}^{\infty}\left(r_{\mathrm{h\_h}}^2\right)\left[\frac{(\sin\theta_{\mathrm{ca}})^2}{A^2}\right]\left[-\frac{1}{h}[(h^2-1)k_L-1]\right]^2\right\}} \tag{3.56.}$$

where $k_L = \frac{\sum_{h=2}^{\infty}\left(r_{\mathrm{h\_h}}^2\right)\left[\frac{1}{h^2}(h^2-1)\right]}{\sum_{h=2}^{\infty}\left(r_{\mathrm{h\_h}}^2\right)\left[\frac{1}{h^2}(h^2-1)^2\right]}$

3.7 Simulation Study

In order to prove the system performance of co-phase traction power supply with Railway HPQC, the analysis is being verified by PSCAD simulations. The simulation circuit schematics are shown in Fig. 3.15. The traction load is simulated using rectifier RLC load. The load parameters are chosen by a load capacity of 31.5 MVA. This value is chosen based on the practical data of WuQing substation. Details of other parameter designs are discussed below.

Determination of different parameters other than those in compensation device are given below.

The traction load is simulated using a controllable current source with external signal in PSCAD based on 31.5 MVA capacity, Load power factor value of 0.85, and harmonic distribution of practical WuQing substation data in Table 3.2.

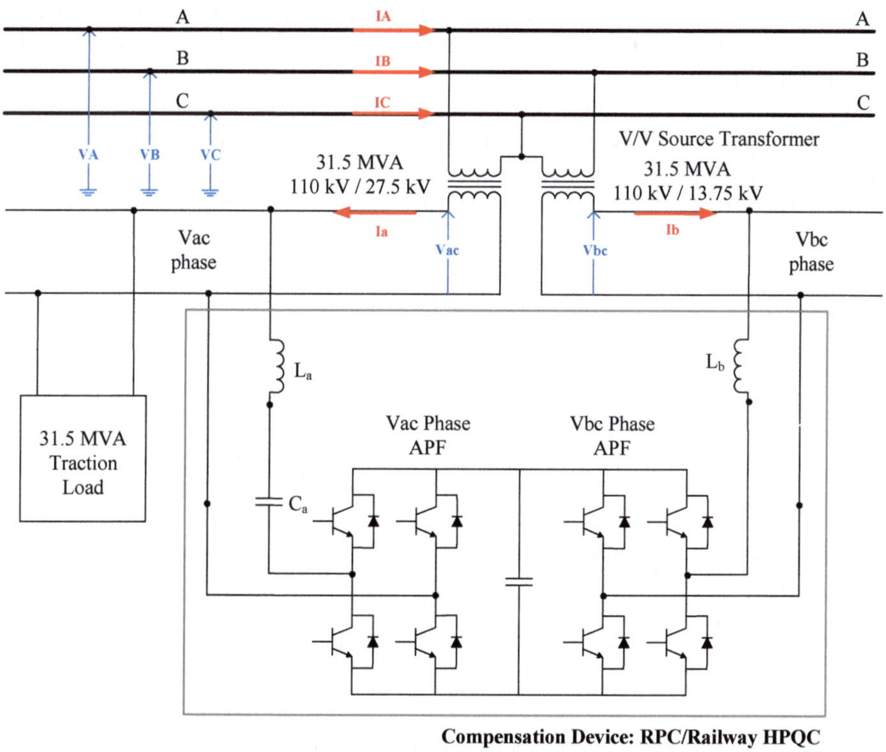

Fig. 3.15 Circuit schematics of co-phase traction power supply with compensation device used in the simulation verifications

3.7.1 System Performance Without Any Compensation

First of all, the system performance without any compensation (neither conventional RPC nor Railway HPQC) is being investigated first. The system source voltage and current are monitored and investigated, and the corresponding waveforms are shown in Fig. 3.16. It can be observed that the system source current suffers from distortion and system unbalance. The phase difference between voltage and current waveforms also signify the presence of reactive power.

3.7.2 System Performance with Conventional RPC Compensation (V_{dc} = 42 kV)

Next, the system performance of co-phase traction power supply with conventional RPC compensation is investigated. The circuit parameters of conventional RPC used are given in Table 3.3. Notice that the dc link voltage used is 42 kV ($k_{\text{invaL}} = 1.07$).

The system source voltage and current waveforms are shown in Fig. 3.17. It could be observed that compared to the waveforms without compensation in Fig. 3.16, the system source current waveform is balanced and harmonic free. The elimination of reactive power can also be investigated by the enhancement of power factor to near unity.

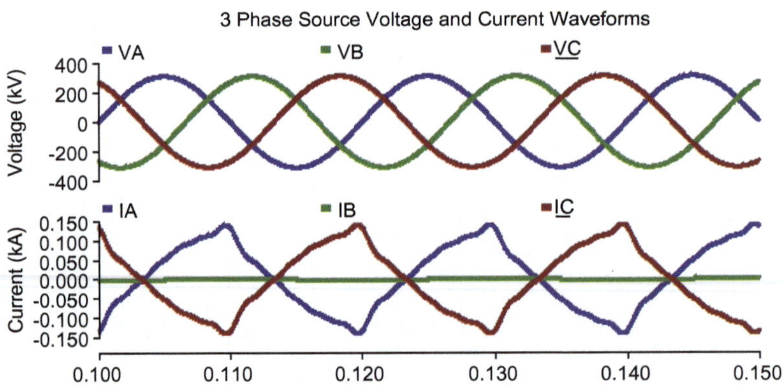

Fig. 3.16 Simulated system source voltage and current waveforms for co-phase traction power supply system without any compensation

Table 3.3 The system parameters used in the simulation of co-phase traction power supply with conventional RPC

No.	Parameters	Symbol	Value
1.	V_{ac} phase coupled impedance ratio	m_{La}	0.074
2.	V_{ac} phase coupled inductance	L_a	6.6 mH
3.	V_{bc} phase coupled inductance	L_b	6 mH
4.	dc link capacitance	C_{dc}	10000 uF
5.	RPC operation voltage rating	k_{invaL}	1.07
6.	dc link operation voltage	V_{dc}	42 kV

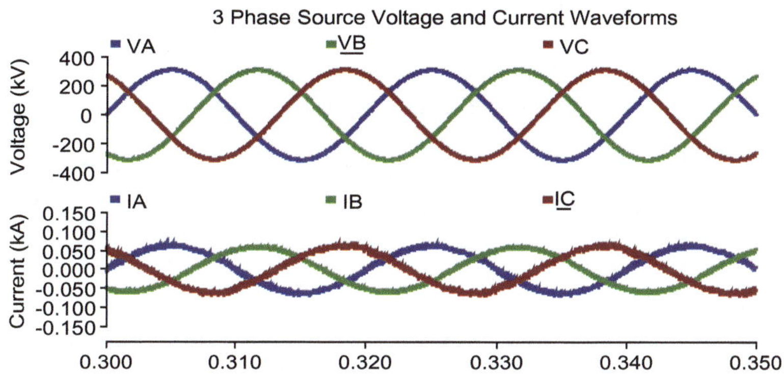

Fig. 3.17 Simulated system source voltage and current waveforms for co-phase traction power supply with Conventional RPC Compensation (V_{dc} = 41.8 kV)

3.7.3 System Performance with Railway HPQC Compensation (Traditional HAPF Design) (V_{dc} = 20 kV)

After simulations of co-phase traction power with conventional RPC, the system performance of co-phase traction power with Railway HPQC is also investigated. First of all, for the harmonic compensation concern, Railway HPQC using traditional HAPF design is being explored first. The circuit parameters of co-phase traction power with Railway HPQC under traditional HAPF design are shown in Table 3.4. The dc link voltage used is 20 kV.

The simulated system source voltage and current waveforms are shown in Fig. 3.18. Comparing the figure with Figs. 3.16 and 3.17, it can be found that the system performance is enhanced with harmonic free and balanced system source current. The phase difference between voltage and current is also less, which indicates elimination of reactive power. The compensation performance is more or less the same as conventional RPC ones in Fig. 3.17. However, there is a total reduction of 52% in the dc link voltage (from RPC: 41.8 kV to Railway HPQC (traditional HAPF design): 20 kV).

Table 3.4 The system parameters used in the simulation of co-phase traction power supply with Railway HPQC using traditional HAPF design

No.	Parameters	Symbol	Value
1.	V_{ac} phase coupled impedance ratio	m_{LCa}	0.8761
2.	V_{ac} phase coupled inductance ratio	k_L	0.125
3.	V_{ac} phase coupled inductance	L_a	9.8 mH
4.	V_{ac} phase coupled capacitance	C_a	110 uF
5.	V_{bc} phase coupled inductance	L_b	6 mH
6.	dc link capacitance	C_{dc}	10000 uF
7.	Railway HPQC operation voltage rating	k_{invaLC}	0.50
8.	dc link operation voltage	V_{dc}	20 kV

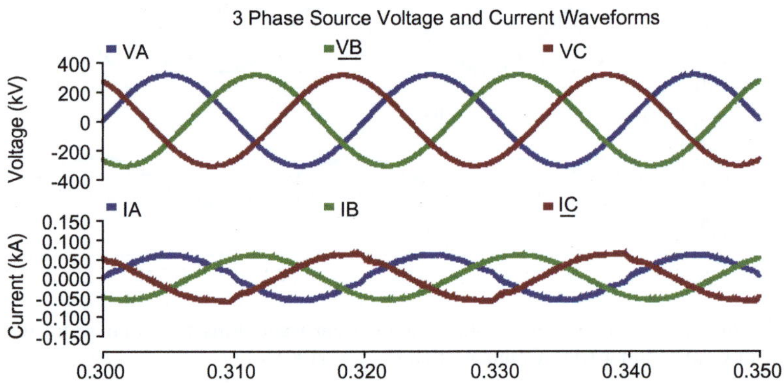

Fig. 3.18 Simulated system source voltage and current waveforms for co-phase traction power supply with Railway HPQC under traditional HAPF design (V_{dc} = 20 kV)

3.7.4 System Performance with Railway HPQC Compensation (New LC Filter Design) (V_{dc} = 18.7 kV)

Finally, the system performance of co-phase traction power supply system with Railway HPQC compensation using new LC filter design). The circuit parameters of Railway HPQC under new LC filter design is shown in Table 3.5. Notice that the dc link voltage required is only 18.7 kV.

Similarly, the simulated waveforms of source voltage and current are shown in Fig. 3.19. Comparing it with Figs. 3.16, 3.17, and 3.18, it can be seen that the system performance is enhanced similar to conventional RPC or Railway HPQC with traditional filter design. However, compared to conventional RPC, there is a reduction of 55% in dc link voltage (from RPC: 41.8 kV to Railway HPQC (new

Table 3.5 The system parameters used in the simulation of co-phase traction power supply with Railway HPQC using new LC filter design

No.	Parameters	Symbol	Value
1.	V_{ac} phase coupled impedance ratio	m_{LCa}	0.074
2.	V_{ac} phase coupled inductance ratio	k_L	0.042
3.	V_{ac} phase coupled inductance	L_a	3.3 mH
4.	V_{ac} phase coupled capacitance	C_a	124 uF
5.	V_{bc} phase coupled inductance	L_b	6 mH
6.	dc link capacitance	C_{dc}	10000 uF
7.	Railway HPQC operation voltage rating	k_{invaLC}	0.4829
8.	dc link operation voltage	V_{dc}	18.7 kV

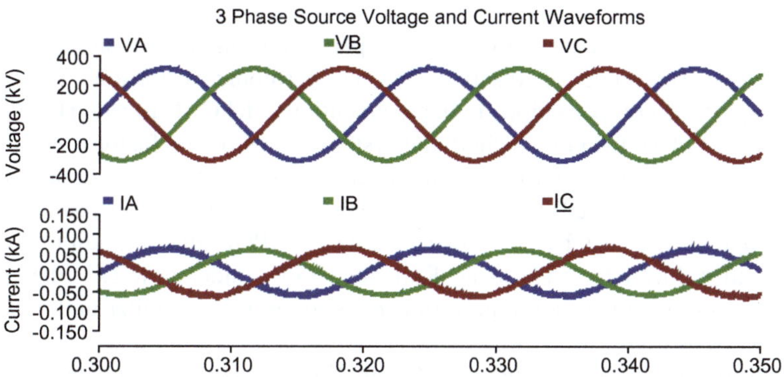

Fig. 3.19 Simulated system source voltage and current waveforms for co-phase traction power supply with Railway HPQC under new LC filter design (V_{dc} = 18.7 kV)

LC filter design): 18.7 kV); compared to Railway HPQC with traditional HAPF design, there is a 6.5% reduction in dc link voltage (from Railway HPQC (traditional HAPF design): 20 kV to Railway HPQC (new LC filter design): 18.7 kV).

3.7.5 Simulation Summary and Further Comparison

In order to have deeper investigation into the system performance of the simulated conditions above. The corresponding monitored system source power factor PF, source current total harmonic distortions (THD) and current unbalance (I_{un}) are shown in the contents below.

System Performance Data Comparison (Satisfactory Compensation)

First of all, the system power quality data of the simulated conditions above are shown in Table 3.6.

Table 3.6 Simulated power quality data of different simulated conditions (satisfactory compensation)

	Three phase source power factor (PF)	Source current total harmonics distortions (THD)	System unbalance (I_{un})	Reduction in V_{dc} (compared to RPC)
Without Compensation	0.60	14.73	99	NA
Conventional RPC (V_{dc} = 41.8 kV)	0.99	4.12	1.74	NA
Railway HPQC (traditional HAPF design) (V_{dc} = 20 kV)	0.99	2.47	3.52	52%
Railway HPQC (new method) (V_{dc} = 18.7 kV)	0.99	2.20	2.99	55.2%

It can be observed that without any compensation, the system suffers from low power factor (presence of reactive power), harmonic distortions and system unbalance. With compensation, either conventional RPC, Railway HPQC using traditional HAPF or the new method, the system source power factor is enhanced to near unity, source current total harmonic distortions and system source current unbalance is also reduced. Compared to conventional RPC compensation, the dc link voltage is reduced by 52% using Railway HPQC under traditional HAPF design, and reduced by 55% using Railway HPQC under new LC filter design, while the system compensation performance is more or less similar under the same rated load condition.

Therefore, it can be concluded that with Railway HPQC using new LC filter design method, there is a maximum of 55% reduction in the dc link operation voltage compared to conventional RPC compensation. Notice that the compensation performance is the same underrated designed load condition.

System Performance Data Comparison (V_{dc} = 18.7 kV)

Next, in order to further verify that the system compensation performance is far from satisfactory with low dc link voltage as in Railway HPQC system, the simulations are performed again with all compensator dc link operation voltage at 18.7 kV. The performance data are shown in Table 3.7.

Table 3.7 Simulated power quality data of different simulated conditions (V_{dc} = 18.7 kV)

	Three phase source power factor (PF)	Source current total harmonics distortions (THD)	System unbalance (I_{un})
Without Compensation	0.60	14.73	99
Conventional RPC (V_{dc} = 18.7 kV)	0.48	10.09	94.03
Railway HPQC (traditional HAPF design) (V_{dc} = 18.7 kV)	0.99	8.95	3.79
Railway HPQC (new method) (V_{dc} = 18.7 kV)	0.99	2.20	2.99

It can be observed that for conventional RPC or Railway HPQC using traditional HAPF design, the system performance is not satisfactory. It is only when the Railway HPQC is designed using new method, satisfactory compensation performance can be provided with a low dc link voltage. This proves the effectiveness of the Railway HPQC is reducing the dc link operation voltage.

3.8 Experimental Verifications

In order to verify the effectiveness of new method for providing satisfactory power quality compensation using Railway HPQC in co-phase traction power with minimum operation voltage under rated loading condition, an experimental hardware prototype is constructed and experimental results are obtained. Details of the experimental setup can be found in 0. The three-phase source voltage and current waveforms, as well as power quality data is monitored using DL750 Oscilloscope and Fluke 430 Power Quality Analyzer.

Similar to simulation verifications, experimental results are obtained under four different conditions: (a) without any compensation device; (b) with conventional RPC compensation (V_{dc} = 80 V); (c) with novel Railway HPQC compensation (traditional harmonic filter design); with novel Railway HPQC compensation (new harmonic filter design).

3.8.1 System Performance Without Any Compensation Device

First of all, the simulated system source voltage and current, as well as load current waveforms of co-phase traction power without any compensation devices are shown in Fig. 3.20. It can be observed that there is high unbalance in the system source current, nearly 100%, and high amount of reactive power and harmonics are present. The system performance is, therefore, far from satisfactory.

The system source is also monitored using Power Quality Analyzer. The screens captured are shown in Fig. 3.21. It could be seen that for the investigated condition of co-phase traction power without any compensation, the system source power factor is 0.70, while the source current harmonics total harmonic distortions (THD) is 28%, and the source current unbalance is 92.7%. These show that for co-phase traction power supply system without any compensation device, it suffers from high presence of reactive power, harmonics and system unbalance. Therefore, power quality compensation is required.

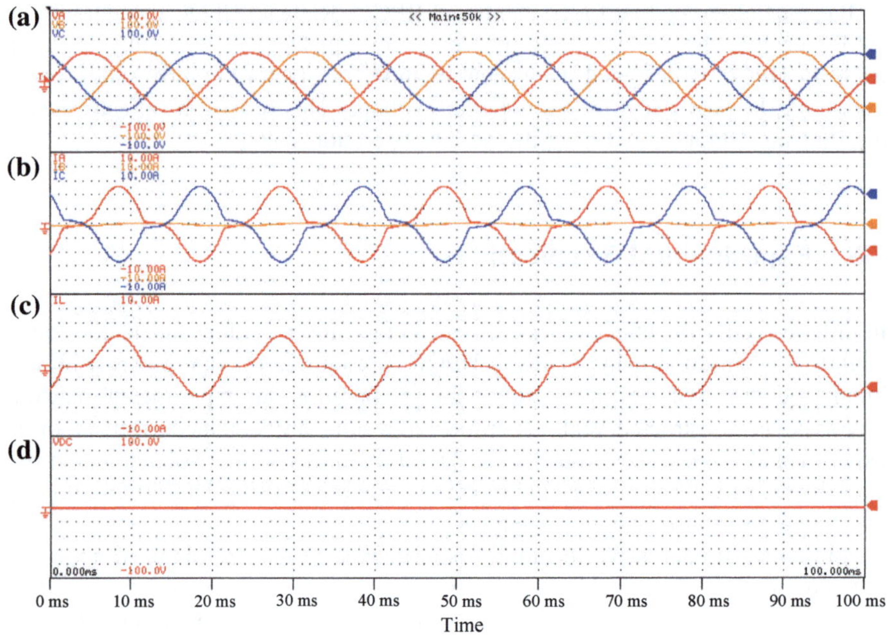

Fig. 3.20 Experimental waveforms obtained from co-phase traction power supply hardware prototype without any compensation: **a** source voltage; **b** source current; **c** traction load current; **d** dc link operation voltage

3.8.2 System Performance with Conventional RPC Compensation (V_{dc} = 80 V)

Next, the system performance of co-phase traction power with conventional RPC compensation (V_{dc} = 80 V) is investigated. Detailed parameter design of conventional RPC is shown in Table 3.8.

The value of coupled inductance ratio is calculated according to (3.8) with current ripple $\Delta I_{ca} = 0.2$ A, and PWM switching frequency $f_s = 10$ kHz. The coupled inductance L_a and operation voltage rating requirement k_{invaL} may then be determined using (3.7) and (3.4) as well as (3.19). Experimental waveforms obtained are shown in Fig. 3.22. Compared to the figure obtained without compensation device in Fig. 3.20, the three-phase source current is more balanced and the harmonic distortion is reduced.

Similarly, the screens obtained from Power Quality Analyzer for co-phase traction power supply with conventional RPC compensation (V_{dc} = 80 V) are presented in Fig. 3.23. It can be observed that the system source power factor is increased from 0.7 to 0.96, and the source current total harmonic distortions (THD) is reduced from 28 to 12.2%, while the system source current unbalance is reduced from 92.7 to 17.3%. However, the dc link voltage used is 80 V, which must be higher than the peak value of PCC voltage (50 V rms, 71 V peak).

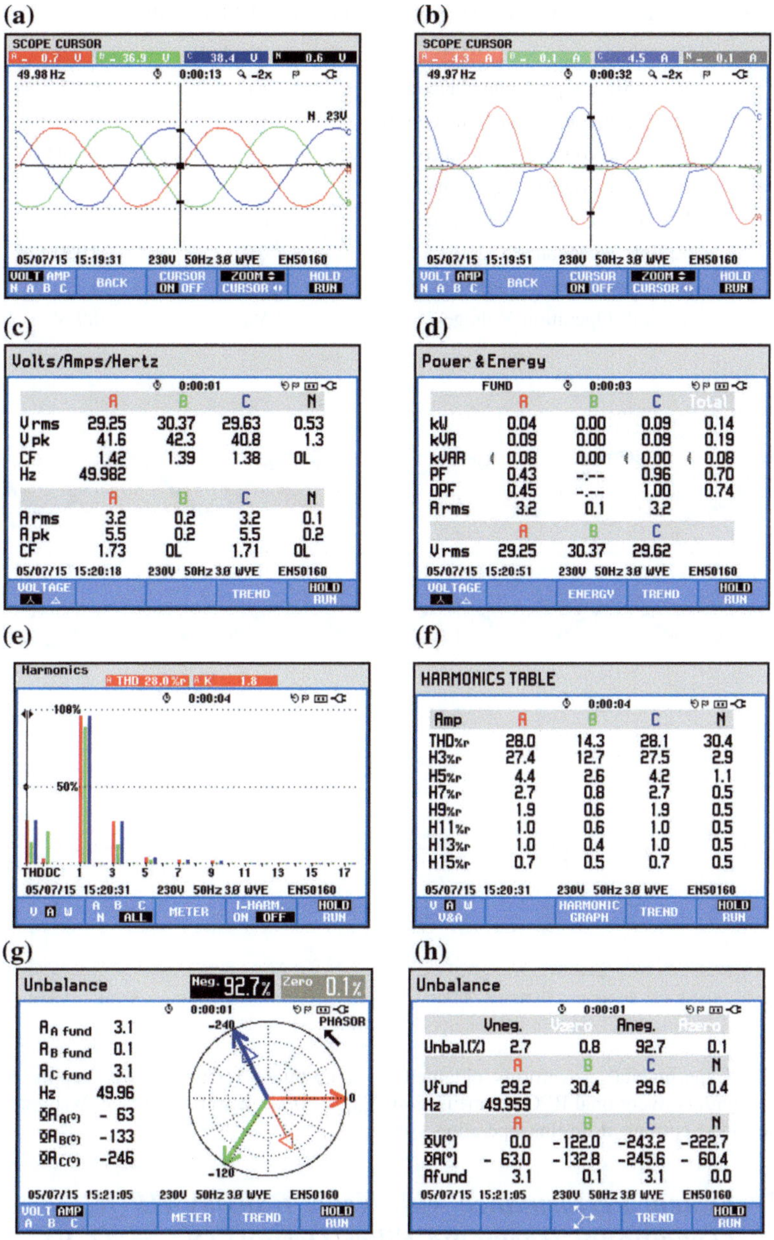

Fig. 3.21 Experimental results of screens captured from hardware prototype of co-phase traction power without any compensation: **a** source voltage waveform; **b** source current waveform; **c** three-phase source voltage and current statistics; **d** three-phase source power and power factor; **e** source current harmonics spectrum; **f** three-phase source current harmonics statistics; **g** source current negative sequence; **h** source voltage and current unbalance data

Table 3.8 Experimental parameter design of conventional RPC for verification

No.	Parameters	Symbol	Value
1.	V_{ac} phase coupled impedance ratio	m_{La}	0.074
2.	V_{ac} phase coupled inductance ratio	k_L	NA
3.	V_{ac} phase coupled inductance	L_a	4 mH
4.	V_{ac} phase coupled capacitance	C_a	NA
5.	V_{bc} phase coupled inductance	L_b	4 mH
6.	dc link capacitance	C_{dc}	10000 uF
7.	RPC Operation Voltage Rating	k_{invaL}	1.13
8.	dc Link Operation Voltage	V_{dc}	80 V

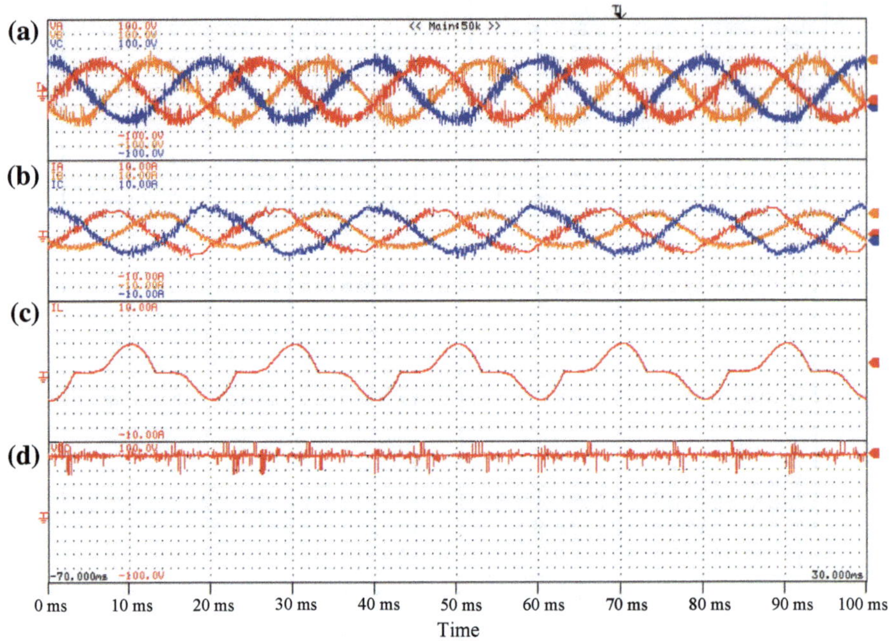

Fig. 3.22 Experimental waveforms obtained from co-phase traction power supply hardware prototype with conventional RPC compensation (V_{dc} = 80 V): **a** source voltage; **b** source current; **c** traction load current; **d** dc link operation voltage

3.8.3 *System Performance with Novel Railway HPQC (Traditional Harmonic Filter Design) (V_{dc} = 42 V)*

Next, the system performance of co-phase traction power supply system with Railway HPQC of traditional harmonic filter design (V_{dc} = 49 V) is further investigated. The experimental parameter design of new Railway HPQC is shown in Table 3.9. The value of coupled impedance is determined according to the design

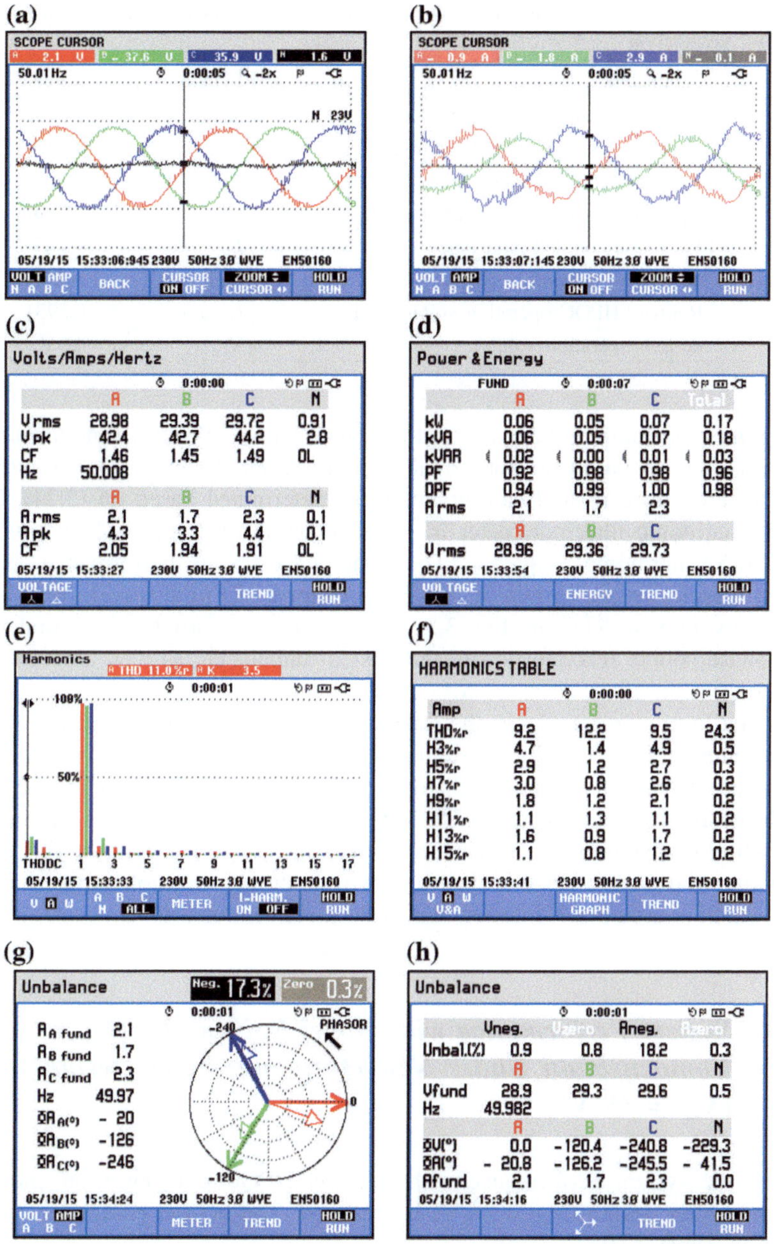

Fig. 3.23 Experimental results of screens captured from hardware prototype of co-phase traction power with conventional RPC compensation (V_{dc} = 80 V): **a** source voltage waveform; **b** source current waveform; **c** three-phase source voltage and current statistics; **d** three-phase source power and power factor; **e** source current harmonics spectrum; **f** three-phase source current harmonics statistics; **g** source current negative sequence; **h** source voltage and current unbalance data

Table 3.9 Experimental parameter design of new railway HPQC under traditional harmonic filter design for verification

No.	Parameters	Symbol	Value
1.	V_{ac} phase coupled impedance ratio	m_{LCa}	0.876
2.	V_{ac} phase coupled inductance ratio	k_L	0.042
3.	V_{ac} phase coupled inductance	L_a	2 mH
4.	V_{ac} phase coupled capacitance	C_a	170 uF
5.	V_{bc} phase coupled inductance	L_b	4 mH
6.	dc link capacitance	C_{dc}	10000 uF
7.	Railway HPQC operation voltage rating	k_{invaLC}	0.5931
8.	dc link operation voltage	V_{dc}	42 V

of minimum operation voltage at rated load in (3.11). The coupled inductance ratio is then computed based on traditional harmonic filter design in (3.39), while the coupled inductance L_a and capacitance C_a is determined based on (3.24). Notice that the harmonic filter parameter is designed based on traditional one, which is tuned at 5th harmonic frequency. The three-phase system performance and load current waveforms are shown in Fig. 3.24. Compared to the waveforms obtained from conventional RPC in Fig. 3.22, the system performance is similar. The three-phase source reactive power and system unbalance, as well as current harmonics are reduced.

Similar to previous analysis, the screens obtained from Power Quality Analyzer are presented in Fig. 3.25. The system source power factor is increased to 0.96, and the source current total harmonic distortions are reduced to 7.9%, while the system unbalance is also reduced to 21.8%. This shows the effectiveness of providing satisfactory compensation performance using Railway HPQC with dc link voltage lower than the peak of PCC voltage. However, the dc link voltage used is 42 V only, which is around 48% lower than that of conventional RPC.

3.8.4 System Performance with Novel Railway HPQC Compensation Under New Harmonic Filter Design (V_{dc} = 38 V)

Finally, the system performance with Railway HPQC compensation under new harmonic filter design (V_{dc} = 38 V) is explored. The system parameter design is shown in Table 3.10.

The coupled impedance X_{LCa} is also determined according to minimum operation voltage at rated load design in (3.11). The coupled inductance ratio k_L is then computed based on the load harmonic distribution, as in (3.45), while the coupled inductance L_a and capacitance C_a can be determined by (3.46) and (3.47). Detailed

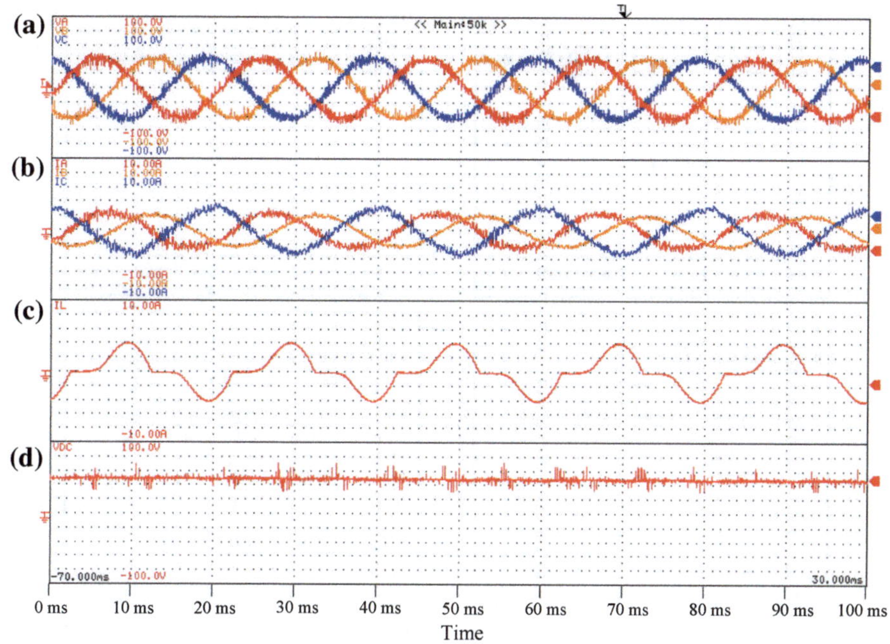

Fig. 3.24 Experimental waveforms obtained from co-phase traction power supply hardware prototype with the Railway HPQC compensation under traditional harmonic filter design (V_{dc} = 42 V): **a** source voltage; **b** source current; **c** traction load current; **d** dc link operation voltage

design procedures can be found in Sect. 3.6. The system source and load waveforms are shown in Fig. 3.26.

Comparing the waveforms with those in Figs. 3.20, 3.22, and 3.24. It can be concluded that the system performance is improved over system without any compensation device, and the compensation performance is similar to conventional RPC and Railway HPQC under traditional harmonic filter design. However, the dc link voltage used is 38 V only.

Similarly, the screens obtained by Power Quality Analyzer for co-phase traction power with the Railway HPQC under new harmonic filter design are presented in Fig. 3.27. It can be seen that the three-phase source power factor is increased to 0.98, and the source current total harmonics is reduced to 6.7%. Moreover, the system source current unbalance is reduced to 22.2%. The compensation performance is similar to conventional RPC and novel Railway HPQC under traditional harmonic filter design. However, by using Railway HPQC under new harmonic filter design, the dc link voltage is reduced by around 50% compared to conventional RPC, and 22% reduction compared to Railway HPQC under traditional harmonic filter design. This shows the validity of Railway HPQC design in minimizing the dc link operation voltage and cost.

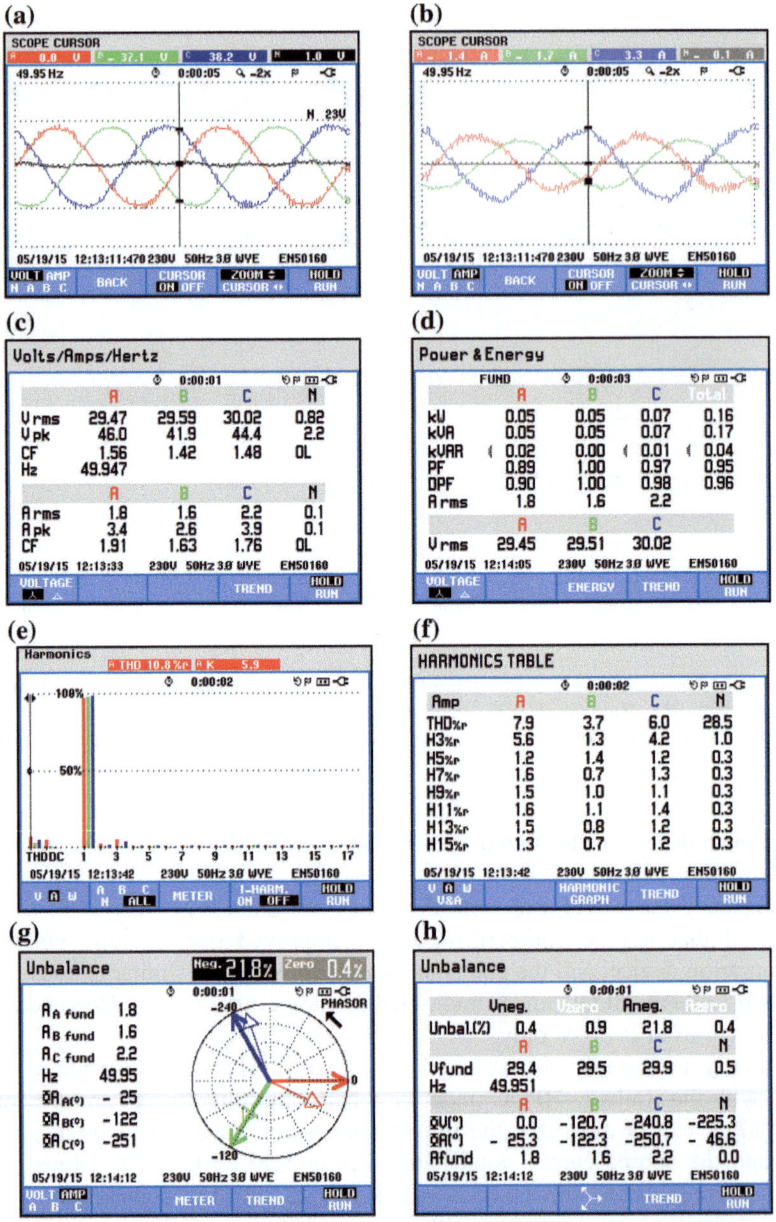

Fig. 3.25 Experimental results of screens captured from hardware prototype of co-phase traction power with novel Railway HPQC compensation under traditional harmonic filter design (V_{dc} = 49 V): **a** source voltage waveform; **b** source current waveform; **c** three-phase source voltage and current statistics; **d** three-phase source power and power factor; **e** source current harmonics spectrum; **f** three-phase source current harmonics statistics; **g** source current negative sequence; **h** source voltage and current unbalance data

Table 3.10 Experimental parameter design of railway HPQC under traditional harmonic filter design for verification

No.	Parameters	Symbol	Value
1.	V_{ac} phase coupled impedance ratio	m_{LCa}	0.876
2.	V_{ac} phase coupled inductance ratio	k_L	0.1052
3.	V_{ac} phase coupled inductance	L_a	4.9 mH
4.	V_{ac} phase coupled capacitance	C_a	150 uF
5.	V_{bc} phase coupled inductance	L_b	4 mH
6.	dc link capacitance	C_{dc}	10000 uF
7.	Railway HPQC operation voltage rating	k_{invaLC}	0.54
8.	dc link operation voltage	V_{dc}	38 V

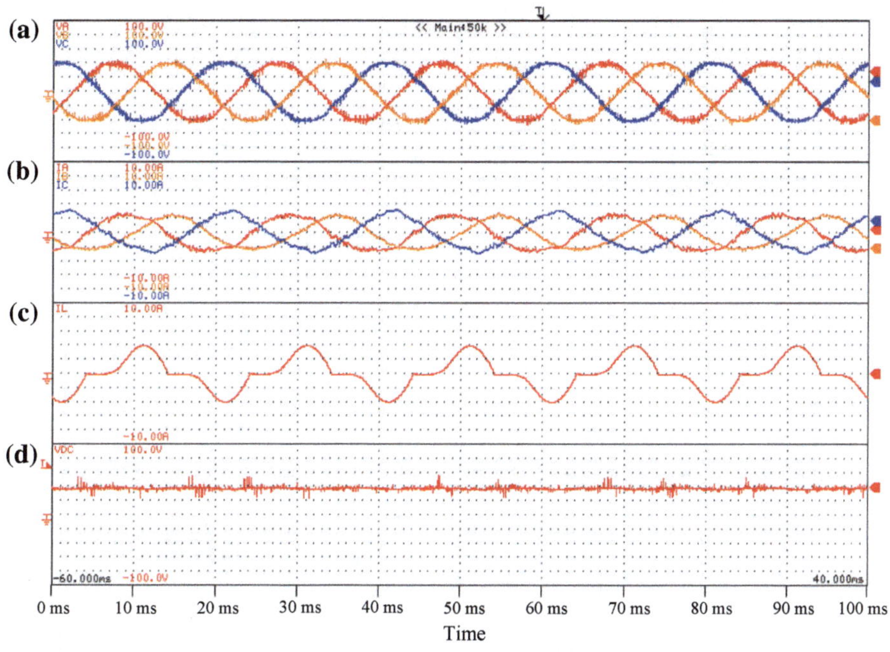

Fig. 3.26 Experimental waveforms obtained from co-phase traction power supply hardware prototype with the Railway compensation under new harmonic filter design (V_{dc} = 38 V): **a** source voltage; **b** source current; **c** traction load current; **d** dc link operation voltage

The system source power quality performance of all simulated conditions above are summarized in Table 3.11.

The following important points can be concluded from the data:

- The system power quality performance is far from satisfactory without any compensation;

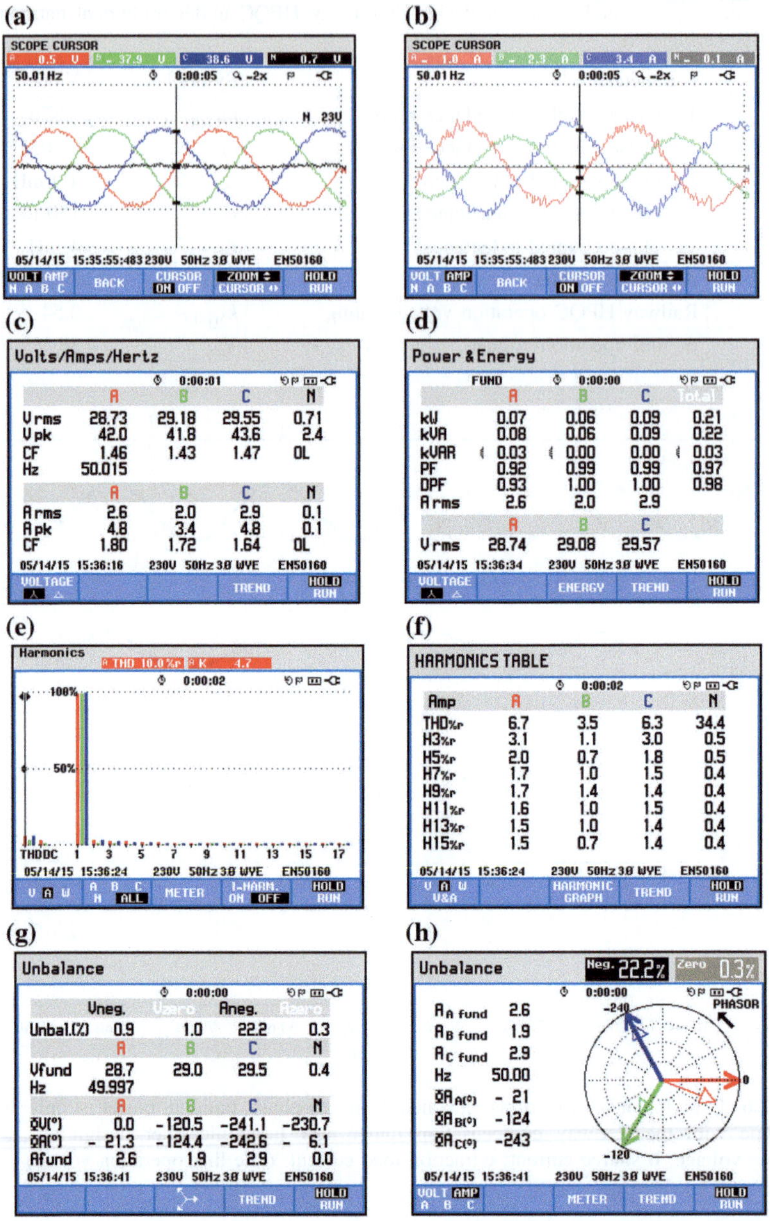

Fig. 3.27 Experimental results of screens captured from hardware prototype of co-phase traction power with Railway HPQC compensation under new harmonic filter design (V_{dc} = 38 V): **a** source voltage waveform; **b** source current waveform; **c** three-phase source voltage and current statistics; **d** three-phase source power and power factor; **e** source current harmonics spectrum; **f** three-phase source current harmonics statistics; **g** source current negative sequence; **h** source voltage and current unbalance data

Table 3.11 Experimental power quality data of different investigated conditions (satisfactory compensation)

	Three phase source power factor (PF)	Source current total harmonics distortions (THD)	System unbalance (I_{un})	Reduction in V_{dc} (compared to RPC)
Without compensation	0.70	28.1	92.7	NA
Conventional RPC (V_{dc} = 80 V)	0.98	12.2	17.3	NA
Railway HPQC (traditional HAPF design) (V_{dc} = 42 V)	0.96	7.9	21.8	47.5%
Railway HPQC (new method) (V_{dc} = 38 V)	0.98	6.7	22.2	52.5%

- The power quality performance is improved and is similar using conventional RPC, Railway HPQC under traditional harmonic filter design, and Railway HPQC under new harmonic filter design;
- However, compared to conventional RPC, the dc link operation voltage is reduced by 47.5% using Railway HPQC with traditional harmonic filter design), and by 52.5% using Railway HPQC with new harmonic filter design.

Therefore, it is shown that the dc link operation voltage can be effectively reduced using novel Railway HPQC with new design.

3.9 Summary

In this chapter, the design of co-phase traction power with Railway HPQC under fixed rated locomotive load condition based on minimum operation voltage requirement is being presented and investigated. The investigations are performed under two different models: fundamental and harmonic ones. A figure of merit, coupled impedance ratio, m_{LCa}, defined as the ratio between coupled impedance (X_{LCa}) and rated compensation impedance (V_{ac}/I_{ca_rated}), is introduced in the analysis.

For fundamental harmonic compensation, it is found that the Railway HPQC can operate at minimum operation voltage when the output compensation current is in phase with the inverter output voltage ($m_{LCa} = \sin\theta_{ca}$).

For the harmonic compensation, a new method different from traditional design is proposed for further minimization of operation voltage, device rating and cost. The harmonic filter is composed of a series inductor and capacitor (LC) branch. Traditionally, the resonant frequency is tuned to the frequency which load harmonics are mostly dominated (like 3rd or 5th harmonic frequency) to reduce the required operation voltage. However, in this case, the operation voltage is not minimized since the load harmonic content distribution spectrum is not considered. Therefore, a new method is developed to determine the coupled capacitance and inductance of the LC branch according to the load harmonic content distribution. The load harmonic spectrum is modeled and two parameters coupled inductance ratio ($k_L = X_L/X_{LCa}$) and coupled capacitance ratio (k_C/X_{LCa}) are added.

Detailed design flow chart of the co-phase traction power with Railway HPQC for fixed rated load under minimum operation voltage requirement is shown in Fig. 3.28. In short, the value of X_{LCa} is determined based on fundamental compensation power requirement while the values of X_L and X_{LCa} (with equivalent impedance of X_{LCa}) are determined based on load harmonic distribution. It is also verified through simulation and experiment that under fixed rated load, with similar system performance, using system with Railway HPQC under new harmonic filter design, the operation voltage can be reduced by around 55% compared to conventional system with RPC. This will also lead to around 55% reduction in device rating, operation loss, and cost.

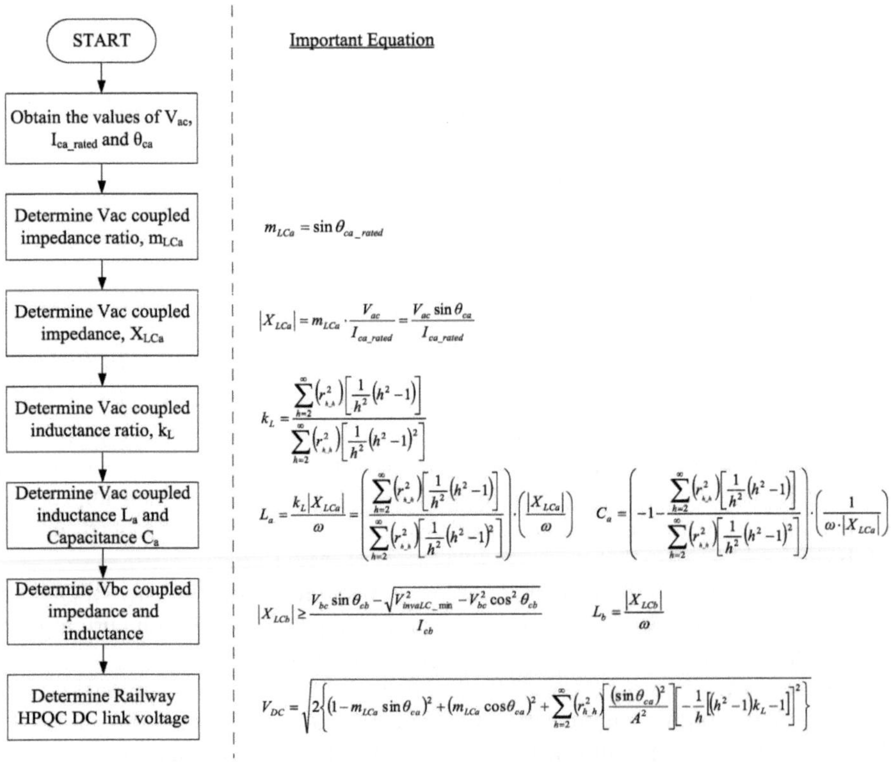

Fig. 3.28 Summarized design flowchart of Railway HPQC in co-phase traction power supply for minimum operation voltage requirement under fixed rated load condition

References

1. Z. Shu, L. Xie and Q. Li, "Single-phase back-to-back converter for active power balancing, reactive power compensation and harmonic filtering in traction power system", IEEE Trans. Power Electronics, Issue 99, 2010.
2. Zeliang Shu; Shaofeng Xie; and Qun-zhan Li, "Development and Implementation of a Prototype for Co-phase Traction Power Supply System", *Asia-Pacific Power and Energy Engineering Conference (APPEEC),* 2010, pp: 1–4.
3. Minwu Chen; Qunzhan Li; and Guang Wei, "Optimized Design and Performance Evaluation of New Cophase Traction Power Supply System", *Asia-Pacific Power and Energy Engineering Conference (APPEEC),* 2009, pp: 1–6.
4. Dai, N.Y.; Lao, K.W.; Wong, M.C.; Wong, C.K., "Hybrid power quality conditioner for co-phase power supply system in electrified railway", *Power Electronics, IET*, 2012, vol: 5, issue: 7, pp: 1084–1094.
5. Lao, K.-W.; Dai, N.; Liu, W.-G.; Wong, M.-C., "Hybrid Power Quality Compensator With Minimum DC Operation Voltage Design for High-Speed Traction Power Systems", *IEEE Transactions on Power Electronics*, 2013, vol: 28, issue: 4, pp: 2024–2036.
6. Ning-Yi Dai; Man-Chung Wong, "Design considerations of coupling inductance for active power filters", *6th IEEE Conference on Industrial Electronics and Applications*, 2011, pp: 1370–1375.

Chapter 4
Various Design Techniques of Co-phase Traction Power with Railway HPQC for Varying Load

Abstract This chapter introduces more flexible design technique for Railway HPQC in co-phase traction power for varying load. In the content, improvement of Railway HPQC performance under varying load is achieved by three different design approaches, namely, raising Railway HPQC operation voltage; redefining the coupled impedance of Railway HPQC by impedance-mapping technique; and having adaptive change in the operation voltage. PSCAD simulation results show the effectiveness of all three approaches. Preliminary experimental results are obtained from laboratory-scaled hardware prototype to show the effectiveness of the design.

4.1 Introduction

In order to achieve a relatively low operation voltage in Railway HPQC when traction load varies, different design techniques can be used. For instance, three possible methods are discussed in this chapter. First, the operation voltage can be raised to provide higher compensation power range so that Railway HPQC can provide satisfactory compensation performance when load varies [1]. The relationship between the compensation range and operation voltage is explored. Second, the power output range of Railway HPQC can be enlarged by redesigning its coupled impedance [2]. The idea works by mapping the coupled impedance with the desired compensation range. The method is thus named as "impedance-mapping" technique. Finally, the operation voltage can be set to vary according to the loading condition, so that a lower operation voltage can be maintained during load variations [3]. The control is being modified in order to achieve variable operation voltage control. Each of the three mentioned methods will be investigated individually in this chapter. Simulation and experimental results are presented to show their performance. A small comparison is also made among them to show their usage for different conditions.

K.-W. Lao et al., *Co-phase Traction Power Supply with Railway Hybrid Power Quality Conditioner*, https://doi.org/10.1007/978-981-13-0438-5_4

4.2 Requirement of Railway HPQC Operation Voltage According to Loading Condition

First of all, it should be noted that the Railway HPQC developed for fixed rated load condition may not be able to provide satisfactory compensation performance when load varies due to the fact that its compensation capability is very narrowed and limited. In order to provide satisfactory compensation under load variations, the operation voltage may need to be enhanced based on the load condition. Therefore, in this section, the load variation and change is modeled first so as to derive the Railway HPQC operation voltage requirement.

4.2.1 Load Variations and Change

In order to do analysis, load variations are modeled first. Load capacity may vary due to different loading conditions. Besides the load capacity changes, the load power factor may also change. Therefore, the load capacity and power factor changes are mainly modeled in this section.

Changes in Load Capacity

The change in locomotive load capacity is mainly modeled by the introduction of parameter r. It is assumed that the load voltage remains constant so that the load capacity is varied such that the load current is changed to r times of rated value, as shown in (4.1).

$$I_L = r \cdot I_{L\_\text{rated}} \tag{4.1}$$

According to (2.17), the expression of I_{ca} can be given by (4.2), and it can be seen that when the load is changed to r times of rated value, the value of I_{ca} is also changed to r times of rated one.

$$\begin{aligned} I_{\text{ca}} &= I_L \cdot \sqrt{(K_1 \text{PF}_L)^2 + \left(K_2 \text{PF}_L + \sqrt{1 - \text{PF}_L^2}\right)^2} \\ &= r \cdot I_{L\_\text{rated}} \cdot \sqrt{(K_1 \text{PF}_L)^2 + \left(K_2 \text{PF}_L + \sqrt{1 - \text{PF}_L^2}\right)^2} \\ &= r \cdot I_{\text{ca\_rated}} \end{aligned} \tag{4.2}$$

It is obvious from (4.2) that when r is greater than 1, the load capacity and compensation current amplitude is larger than rated value; on the other hand, when *r* is lower than 1, the load capacity and compensation current amplitude is lower than rated value. A vector diagram showing the load variations with constant locomotive load power factor affecting the operation of conventional RPC and novel Railway HPQC in co-phase traction power is given in Fig. 4.1.

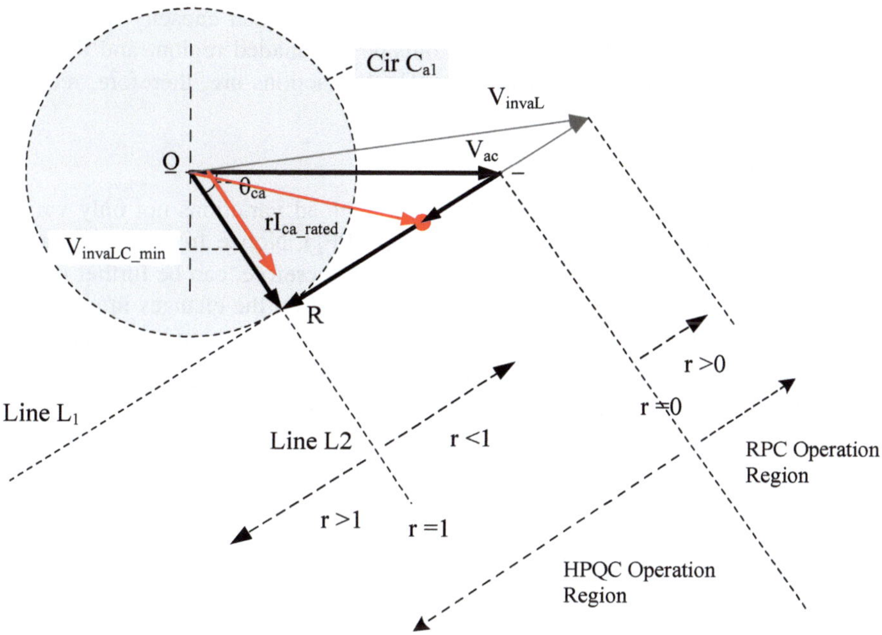

Fig. 4.1 A vector diagram showing the conventional RPC and Railway HPQC condition when load capacity changes (with constant load power factor) in co-phase traction power

Referring to the expression of compensation angle θ_{ca} in (2.47), the value of θ_{ca} is dependent only on the load power factor PF_L. With constant load power factor PF_L, the value of θ_{ca} is also fixed and constant.

Therefore, with constant load power factor, the vector across coupled impedance in conventional RPC V_{La}, and that in novel Railway HPQC V_{LCa}, would vary along the line L_1. In conventional RPC, as r increases, the vector V_{LCa} will grow toward the right-hand side away from the origin. In contrast, in the Railway HPQC, when the system is designed based on rated value according to the contents of Chap. 3, the edge of the vector V_{LCa} would be located at point R, which is perpendicular to the vector of compensation current I_{ca}. When r is larger than rated value 1.0, the amplitude of V_{LCa} (which is product of multiplication between coupled impedance X_{LCa} and compensation current I_{ca}) increases along line L_1 such that the edge of V_{LCa} is located on the left-hand side of the point R; on the other hand, when r is smaller than rated value 1.0, the amplitude of V_{LCa} decreases and the edge of V_{LCa} is located on the left-hand side of point R.

As discussed, when the Railway HPQC in co-phase traction power system is designed for minimum operation voltage at rated design value, the edge of V_{LCa} would be located at point R (perpendicular to vector I_{ca} along line L_1). The satisfactory compensation range of Railway HPQC is then a circle with center at origin and radius of k_{invaLC_min}, shown as shaded area in Fig. 4.1. Satisfactory compensation performance can be provided only when the edge of V_{LCa} is located within this shaded region. Thus, it can be seen that when the load capacity differs from rated value, the edge of V_{LCa} would lie outside the shaded region, and the compensation performance is not satisfactory. Further actions are, therefore, required and will be discussed in this chapter.

Changes in Load Power Factor

It is also assumed that as traction load varies, load variations not only vary in load capacity (r) but also in load power factor (PF_L). Notice from (2.47) that θ_{ca} changes with different load power factor PF_L and, therefore, can be further used to model load variations. The load variation is modeled by the changes in active and reactive power. Assuming that the required V_{ac} phase active and reactive power changes as shown in (14) and (15), the ratio of cos θ_{ca} to cos θ_{ca_rated} is defined as h_c, and the ratio of sin θ_{ca} to sin θ_{ca_rated} is defined as h_s.

$$\frac{p_{ca}}{p_{ca\_rated}} = \frac{r \cdot (I_{ca\_rated}) \cdot (\cos\theta_{ca})}{(I_{ca\_rated}) \cdot (\cos\theta_{ca\_rated})} = r \cdot \left(\frac{\cos\theta_{ca}}{\cos\theta_{ca\_rated}}\right) = r \cdot h_c \tag{4.3}$$

$$\frac{q_{ca}}{q_{ca\_rated}} = \frac{r \cdot (I_{ca\_rated}) \cdot (\sin\theta_{ca})}{(I_{ca\_rated}) \cdot (\sin\theta_{ca\_rated})} = r \cdot \left(\frac{\sin\theta_{ca}}{\sin\theta_{ca\_rated}}\right) = r \cdot h_s \tag{4.4}$$

When the parameters r, h_s, and h_c are included, the expressions of I_{cap} and I can be manipulated as (4.5).

$$\begin{cases} I_{cap} = r \cdot I_{ca\_rated} \cdot h_c \cdot \cos\theta_{ca\_rated} \\ I_{caq} = r \cdot I_{ca\_rated} \cdot h_s \cdot \sin\theta_{ca\_rated} \end{cases} \tag{4.5}$$

The effect of locomotive load power factor variation may also be investigated in another way. According to (2.47), the value of the compensation angle θ_{ca} varies with load power factor, for ideal power quality compensation, $K_1 = 0.5$, $K_2 = 0.2887$, a Matlab plot is constructed and shown in Fig. 4.2. It can be observed from the figure that for inductive traction load of power factor value from 0 to 1, the value of θ_{ca} goes from 90° to 30°.

As shown in Fig. 4.3 is a vector diagram showing the effects on novel Railway HPQC operation when locomotive load power factor PF_L varies. In novel Railway HPQC, since the coupled impedance is capacitive, the vector V_{LCa} is always 90° clockwise the direction of compensation current vector I_{ca}. Therefore, the direction of vector V_{LCa} will vary with changes on locomotive load power factor PF_L. Referring the variation of θ_{ca} with different locomotive load power factor PF_L in Fig. 4.2, it could be observed that when locomotive load power factor PF_L equals zero, the value of θ_{ca} is 90° and the direction of V_{LCa} required would be parallel with the PCC voltage vector V_{ac} but in the opposite direction. As power factor PF_L increases, the value of θ_{ca} decreases, and the direction of the required V_{LCa} will, therefore, rotate anticlockwise until PF_L equals 1.0, and the angle between V_{ac} and V_{LCa} is $90° + \theta_{ca} = 90° + 30° = 120°$. There is no condition that the angle between V_{ac} and V_{LCa} will be smaller than 120°.

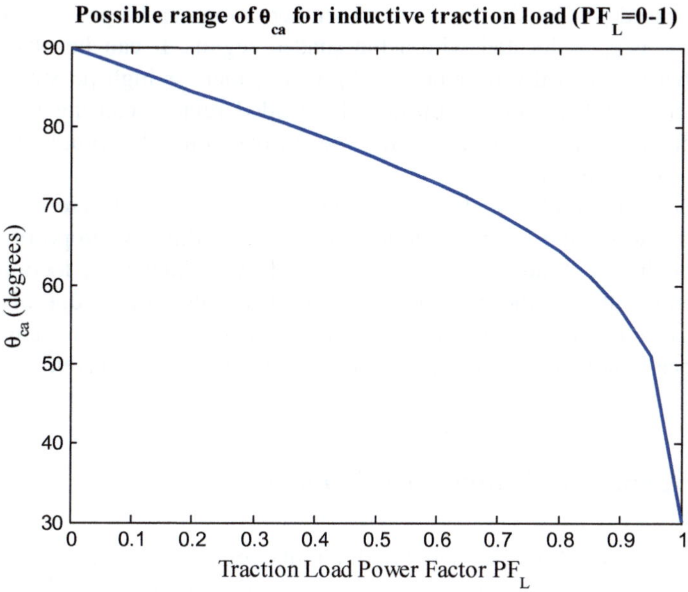

Fig. 4.2 A Matlab plot showing the variation of compensation angle θ_{ca} with different locomotive load power factor PF_L

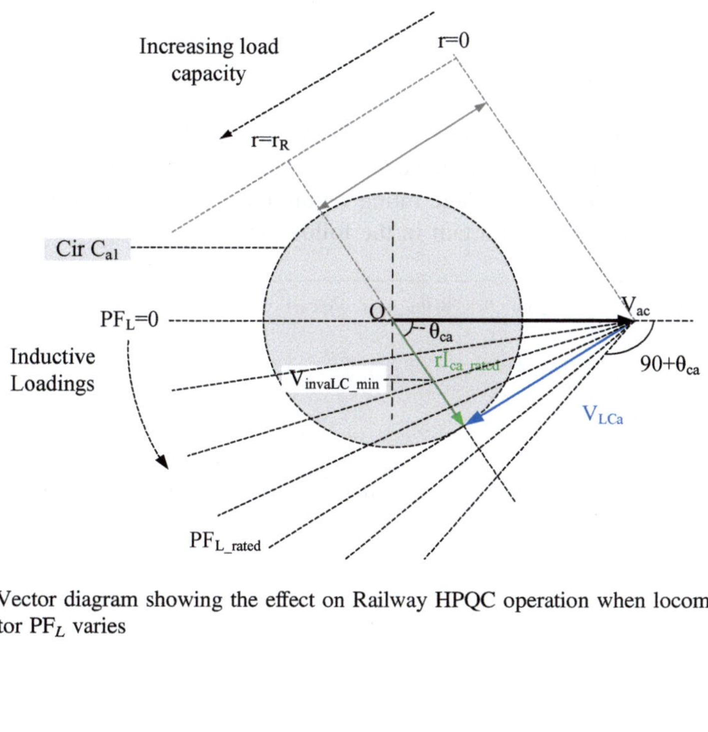

Fig. 4.3 Vector diagram showing the effect on Railway HPQC operation when locomotive load power factor PF_L varies

As discussed previously, satisfactory compensation can be provided only when the edge of V_{LCa} is located within the shaded region. It can be observed from Fig. 4.3 that when load power factor PF_L varies, there is high possibility that the edge of required V_{LCa} locates outside the shaded region, causing compensation performance to be not satisfactory. In order to overcome this problem, actions or modifications are required.

For a complete understanding, the changes in load capacity are also shown in Fig. 4.3. In short, the changes in load capacity is directly proportional to the changes in the amplitude of vector V_{LCa}, while the changes in locomotive load power factor will alter the direction of vector V_{LCa}. When the edge of the vector V_{LCa} is located outside compensation range (shaded region), the compensation performance is not satisfactory and further actions need to be taken.

4.2.2 Operation Voltage Requirement

As can be seen from Figs. 4.1 and 4.3, different operation voltage ratings (for radius of shaded region) are required for different loading conditions. Therefore, it is important to derive the operation voltage requirement first before further analysis.

Referring back to (2.40), the relationship between Railway HPQC operation voltage and locomotive loading condition can be known. This important equation is shown again in (4.6) below.

$$\left|\vec{V}_{\mathrm{invaLC}}\right| = \sqrt{\left(V_{\mathrm{ac}} - I_{\mathrm{caq}} \cdot |X_{\mathrm{LCa}}|\right)^2 + \left(I_{\mathrm{cap}} \cdot |X_{\mathrm{LCa}}|\right)^2} \tag{4.6}$$

By substituting (4.5) into (4.6), the expression of Railway HPQC operation voltage requirement when load changes can be obtained, as shown in (4.7). The Railway HPQC operation voltage rating requirement is then given by (4.8). This expression is especially important in the following analysis.

$$\left|\vec{V}_{\mathrm{invaLC}}\right| = \sqrt{(V_{\mathrm{ac}} - r \cdot I_{\mathrm{ca\_rated}} \cdot h_s \cdot \sin\theta_{\mathrm{ca\_rated}} \cdot |X_{\mathrm{LCa}}|)^2 + (r \cdot I_{\mathrm{ca\_rated}} \cdot h_c \cdot \cos\theta_{\mathrm{ca\_rated}} \cdot |X_{\mathrm{LCa}}|)^2} \tag{4.7}$$

$$\begin{aligned} k_{\mathrm{invaLC}} &= \frac{\left|\vec{V}_{\mathrm{invaLC}}\right|}{V_{\mathrm{ac}}} \\ &= \sqrt{(1 - r \cdot h_s \cdot \sin\theta_{\mathrm{ca\_rated}} \cdot m_{\mathrm{LCa}})^2 + (r \cdot h_c \cdot \cos\theta_{\mathrm{ca\_rated}} \cdot m_{\mathrm{LCa}})^2} \end{aligned} \tag{4.8}$$

4.3 Enhancing Railway HPQC Compensation Capability by Increasing Operation Voltage (Based on Rated Coupled Impedance Design)

The first method to provide satisfactory compensation performance during load variations is to enhance Railway HPQC Compensation Capability by increasing its operation voltage. Detailed idea is shown in this section. Notice that in this method, the V_{ac} coupled impedance of Railway HPQC is kept constant at rated design value as determined by (3.11).

4.3.1 Railway HPQC Operation Voltage Requirement Based on Load Condition Variations

First of all, the parameter definition is as follows. Since in the method of enhancing Railway HPQC compensation capability by increasing operation voltage rating, the V_{ac} coupled impedance is kept constant at design rated value in (3.11) of Chap. 3 based on minimum operation voltage requirement at rated load condition. In order to distinguish, the subscript "_rated" is added to show it is the design under rated load condition, as shown in (4.9).

$$|X_{\mathrm{LCa\_rated}}| = m_{\mathrm{LCa\_rated}} \cdot \frac{V_{\mathrm{ac}}}{I_{\mathrm{ca\_rated}}} = \sin\theta_{\mathrm{ca\_rated}} \cdot \frac{V_{\mathrm{ac}}}{I_{\mathrm{ca\_rated}}} \tag{4.9}$$

By substituting $|X_{\mathrm{LCa}}|$ in (4.7) by $|X_{\mathrm{LCa\_rated}}|$ in (4.9), the expression in (4.10) can be obtained.

$$|\vec{V}_{\mathrm{invaLC}}| = \sqrt{\left(V_{\mathrm{ac}} - r \cdot h_s \cdot (\sin\theta_{\mathrm{ca\_rated}})^2 \cdot V_{\mathrm{ac}}\right)^2 + (r \cdot h_c \cdot \sin\theta_{\mathrm{ca\_rated}} \cdot \cos\theta_{\mathrm{ca\_rated}} \cdot V_{\mathrm{ac}})^2} \tag{4.10}$$

In order to eliminate the effect of PCC voltage during analysis, the expression in (4.10) is divided by V_{ac} so as to obtain the expression for determination of Railway HPQC based on load condition variations in (4.11).

$$k_{\mathrm{invaLC}} = \sqrt{\left(1 - r \cdot h_s \cdot (\sin\theta_{\mathrm{ca\_rated}})^2\right)^2 + (r \cdot h_c \cdot \sin\theta_{\mathrm{ca\_rated}} \cdot \cos\theta_{\mathrm{ca\_rated}})^2} \tag{4.11}$$

4.3.2 Relationship Between Operation Voltage Rating and Compensation Capability

After deriving the Railway HPQC operation voltage requirement based on load condition variation, it can be observed that the Railway HPQC can provide satisfactory compensation performance during load variations when the operation voltage is high enough. However, it is important to determine the appropriate operation voltage rating since too high an operation voltage may lead to high device rating and operation loss, which degrades the advantages of the Railway HPQC.

As shown in Fig. 4.4 is a revision of Fig. 4.1 to show clearly the conventional RPC and novel HPQC operation with enhanced operation voltage to increase compensation capability. It is first supposed that the Railway HPQC Vac coupled impedance is designed based on rated load condition according to (3.11). Under such condition, the minimum Railway HPQC operation voltage V_{invaLC_min}, is always parallel with compensation current I_{ca_rated} and is perpendicular to the vector V_{LCa}. The compensation range is, therefore, bounded by the circle Cir C_{a1}, with radius of V_{invaLC_min} and is shown as a shaded region in Fig. 4.4. As discussed previously,

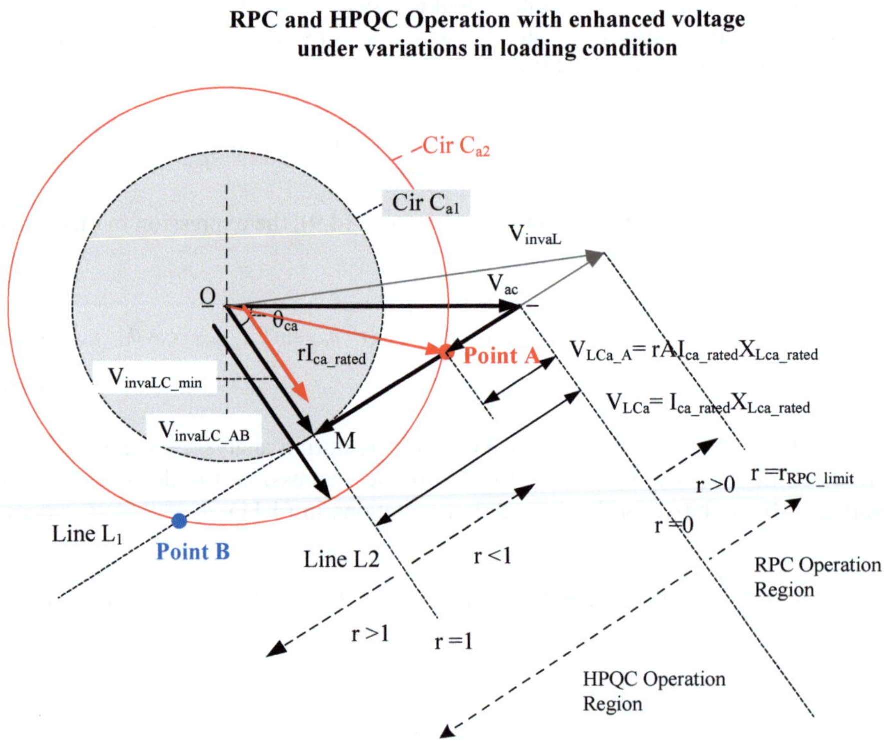

Fig. 4.4 A vector diagram showing the operation of conventional RPC and novel Railway HPQC with enhanced operation voltage rating to increase compensation capability

the compensation performance may not be satisfactory when load varies since the edge of vector V_{LCa} may be located outside the shaded region. Therefore, at minimum Railway HPQC operation voltage, the compensation capability is limited.

By increasing the Railway HPQC operation voltage to be larger than V_{invaLC_min}, the compensation capability can be enhanced. For example, when the operation voltage is increased to the outermost circle Cir C_{a2}, the compensation region can be enlarged so that satisfactory compensation performance can be provided for a larger range of loading conditions.

Conventional RPC Operation Voltage Rating

For comparison, the operation of conventional RPC during load variations is analyzed first. By substituting the coupled inductance ratio m_{La} definition in (2.41) and the load variation model in (4.5) into the conventional RPC operation voltage determination in (2.35), the conventional RPC operation voltage rating in (4.12) can be obtained.

$$k_{invaL\_AB} \geq \sqrt{(1 + r \cdot h_s \cdot m_{La} \cdot \sin\theta_{ca\_rated})^2 + (r \cdot h_c \cdot m_{La} \cdot \cos\theta_{ca\_rated})^2} \tag{4.12}$$

By further manipulation of (4.12), the loading condition limit for conventional RPC operation can be obtained and is determined by the expression of r_{RPC_limit} in (4.13).

$$r_{RPC\_limit} = \frac{-(2 \cdot h_s \cdot m_{La} \cdot \sin\theta_{ca\_rated}) \pm \sqrt{(2 \cdot h_s \cdot m_{La} \cdot \sin\theta_{ca\_rated})^2 - 4(m_{La})^2\left(1 - k_{invaL\_AB}^2\right)}}{2(m_{La})^2} \tag{4.13}$$

For example, according to the computation in Table 3.1, the value of conventional RPC coupled impedance ratio m_{La} is 0.074. With common locomotive load power factor value of 0.85, a Matlab plot is constructed in Fig. 4.5 to show the conventional RPC compensation capability under rated load power factor of 0.85 based on (4.12) and (4.13). The followings can be observed from Fig. 4.5:

- With increased operation voltage rating in conventional RPC, the range of loading conditions that satisfactory compensation performance can be provided is also larger;
- In order to let conventional RPC provide satisfactory compensation performance, the RPC operation voltage rating must be larger than 1.0, which is a lot higher than that in the Railway HPQC.
- Novel Railway HPQC Operation Voltage Rating.

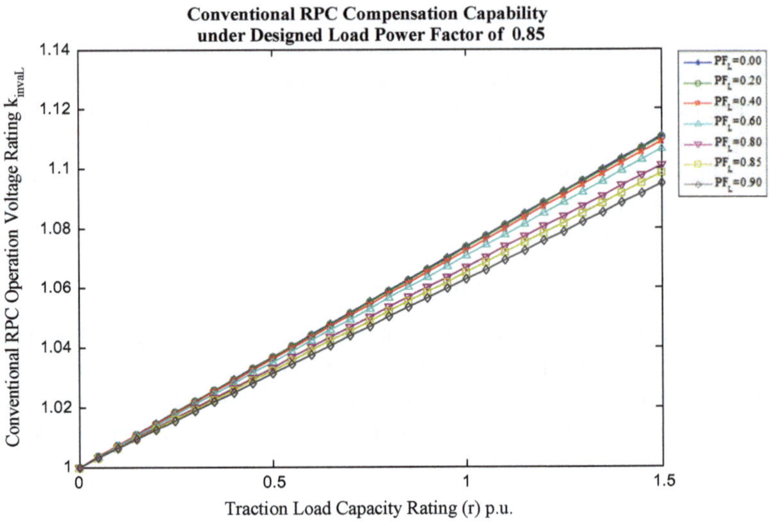

Fig. 4.5 A Matlab plot showing the conventional RPC compensation capability when load power factor varies (RPC designed based on rated load power factor of 0.85)

Based on (4.11), the relationship between Railway HPQC operation voltage rating and compensation capability can be explored. Supposing that the compensation range of load capacity goes from r_A to r_B, the required operation voltage rating is defined as $k_{\text{invaLC\_AB}}$ in (4.14).

$$k_{\text{invaLC\_AB}} \geq \sqrt{\left(1 - r \cdot h_s \cdot (\sin\theta_{\text{ca\_rated}})^2\right)^2 + (r \cdot h_c \cdot \sin\theta_{\text{ca\_rated}} \cdot \cos\theta_{\text{ca\_rated}})^2} \quad (4.14)$$

Solving (4.14) for the value of r, then represents the loading capacity boundaries, the results are shown in (4.15). r_A signifies the lower boundary condition while r_B signifies the upper boundary condition. The value of h_c is not shown in the results since $(h_s)^2(\sin\,\theta_{\text{ca\_rated}})^2 + (h_c)^2(\cos\,\theta_{\text{ca\_rated}})^2 = 1$. The middle point r_M is always located at h_s with range of 2Δ.

$$\begin{aligned} & r_A \leq r \leq r_B \\ & r_A = h_s - \sqrt{(h_s)^2 - \frac{1-(k_{\text{invaLC\_AB}})^2}{(\sin\theta_{\text{ca\_rated}})^2}} = h_s - \Delta \\ & r_B = h_s + \sqrt{(h_s)^2 - \frac{1-(k_{\text{invaLC\_AB}})^2}{(\sin\theta_{\text{ca\_rated}})^2}} = h_s + \Delta \\ & r_M = \frac{r_A + r_B}{2} = h_s \\ & \text{range} = r_B - r_A = 2\Delta \end{aligned} \quad (4.15)$$

A Matlab plot is constructed in Fig. 4.6 to show the novel Railway HPQC compensation capability under rated load power factor of 0.85 based on (4.14) and (4.15). For example, the minimum Railway HPQC computed at $PF_L = 0.85$ is about 0.48. When the value of Railway HPQC operation voltage rating k_{invaLC} is increased to 0.66, the intersections between the horizontal line $k_{invaLC\_AB} = 0.66$ and the curves represent the boundary load capacity points (point *A* for lower boundary; point *B* for upper boundary).

A revision of the vector diagram in Fig. 4.3 is shown in Fig. 4.7 to show the compensation capability of Railway HPQC under designed rated value with increased operation voltage.

- As long as the Railway HPQC operation voltage is higher than its minimum value, it can always provide satisfactory compensation performance at rated condition;
- When the HPQC operation voltage rating is higher, the range of loading conditions that satisfactory compensation performance can be provided (compensation range) also gets larger;
- The middle point r_M is located at $r = 1$ when $PF_L = PF_{L\_rated}$; the value of r_M gets smaller when $PF_L > PF_{L\_rated}$, and vice versa;
- As load power factor PF_L increases, the compensation capability in terms of traction load capacity rating decreases.

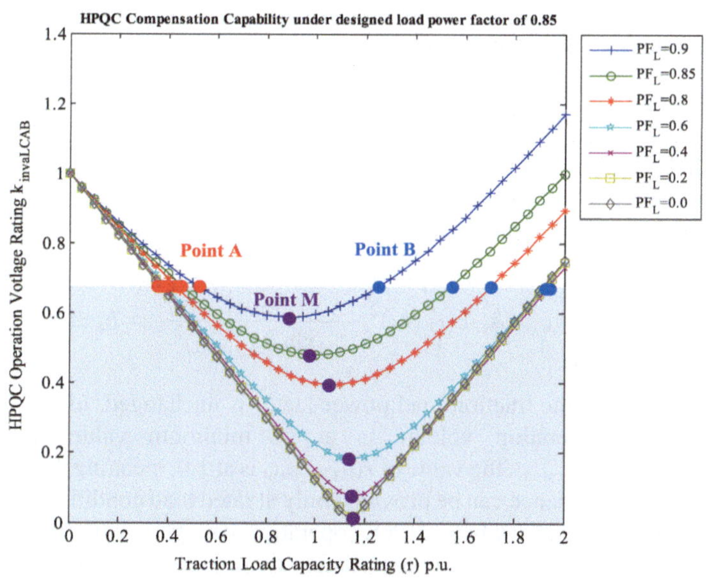

Fig. 4.6 A Matlab plot showing the novel Railway HPQC compensation capability when load power factor varies (HPQC designed based on rated load power factor of 0.85)

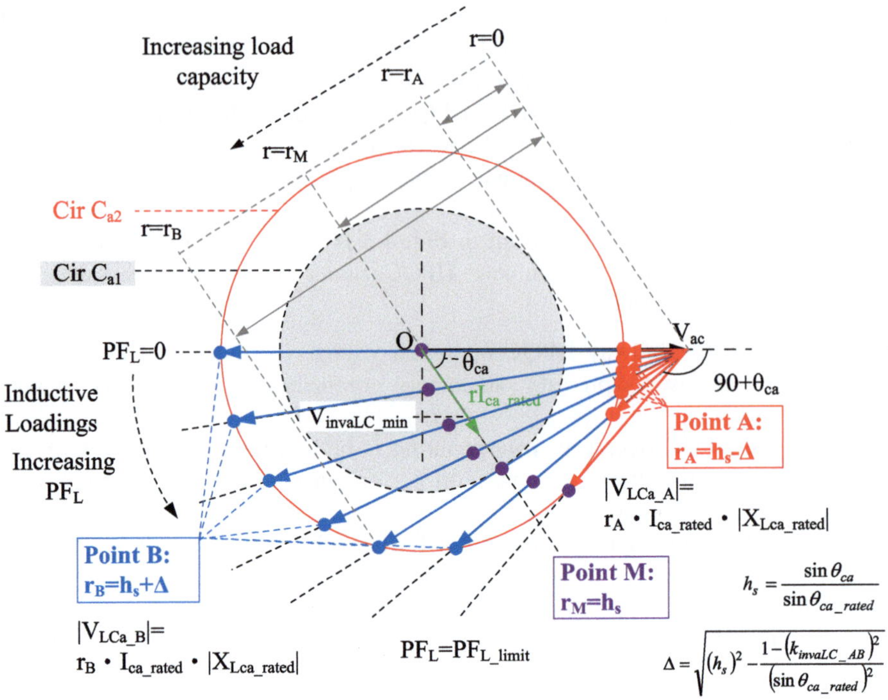

Fig. 4.7 Vector diagram showing the compensation capability of Railway HPQC under designed rated value with increased operation voltage

Limit in Load Capacity Rating

As can be observed from Figs. 4.6 and 4.7, given a certain Railway HPQC operation voltage rating, there is a limit on the load capacity rating. This limit can be deduced from (4.15) and is shown in (4.16).

$$r_{\mathrm{HPQC\_limit}} = h_s \pm \sqrt{(h_s)^2 - \frac{1-(k_{\mathrm{invaLC\_AB}})^2}{(\sin\theta_{\mathrm{ca\_rated}})^2}} = h_s \pm \Delta \tag{4.16}$$

Notice that when the traction load power factor is unchanged, $hs = 1$. When the Railway HPQC operation voltage is at its minimum value as in (3.14) ($k_{\mathrm{invaLC\_AB}} = \cos\theta_{\mathrm{ca\_rated}}$), the value of $r_{\mathrm{HPQC\_limit}}$ is at 1.0, meaning that satisfactory compensation performance can be provided only at rated load condition. On the other hand, when the value of Railway HPQC operation voltage reaches that in conventional RPC ($k_{\mathrm{invaLC\_AB}} = 1$), the compensation range would go from 0 to 2 p.u.

Load Power Factor Limit

Furthermore, the following can be observed from Fig. 4.7:

- As load power factor PF_L increases, the vector V_{LCa} rotates toward the lower half of the circle Cir C_{a2} around the edge of vector V_{ac}.

- With a certain value of $V_{\text{invaLC\_AB}}$, there is a boundary condition when V_{LCa} is tangent to the circle Cir C_{a2}; this is also the limit of load power factor variations, which will be discussed below.

As load power increases, the vector V_{LCa} rotates until a condition that the edge of V_{LCa} is located outside the Railway HPQC compensation region. Under such condition, the Railway HPQC cannot provide satisfactory compensation performance. This happens when there is no intersection point. In short, there is no rational solution of r for the expressions in (4.14). According to the mathematical theory, this refers to the condition when Δ is not rational, as expressed in (4.17).

$$(h_s)^2 - \frac{1-(k_{\text{invaLC\_AB}})^2}{(\sin\theta_{\text{ca\_rated}})^2} < 0 \tag{4.17}$$

By substituting the definition of h_s in (4.4) into (4.17), the expression in (4.18) can be obtained.

$$\cos\theta_{\text{ca}} < k_{\text{invaLC\_AB}} \tag{4.18}$$

Therefore, the HPQC compensation capability is limited by (4.18). Referring back to the definition of θ_{ca}, the power factor limit can then be determined as (4.19). The load power factor limit can then be determined by (4.20). It is interesting that the load power factor limit is not related to any rated value of the Railway HPQC design, but is related to the Railway HPQC operation voltage rating $k_{\text{invaLC\_AB}}$ only.

$$\theta_{\text{ca\_limit}} = \cos^{-1}(k_{\text{invaLC\_AB}}) \tag{4.19}$$

$$\text{PF}_{\text{L\_limit}} = \sqrt{\frac{4}{4+\left[\tan(\cos^{-1}(k_{\text{invaLC\_AB}})) - \frac{1}{\sqrt{3}}\right]^2}} \tag{4.20}$$

The plot of Railway HPQC operation voltage rating $k_{\text{invaLC\_AB}}$ against power factor limit $\text{PF}_{\text{L\_limit}}$ is shown in Fig. 4.8. It is observed that as the Railway HPQC operation voltage rating increases, the power factor limit also increases. For instance, with Railway HPQC operation voltage rating of 0.66, the power factor with HPQC operation voltage rating of 0.66, the power factor limit is somewhere around 0.96. In other words, the HPQC cannot provide satisfactory compensation performance once the traction load power factor exceeds 0.96. The Railway HPQC operation voltage may, therefore, also be chosen based on the desired load power factor limit.

The analysis of relationship between Railway HPQC operation voltage and compensation capability within this section is summarized below.

- Referring to Fig. 4.4, when Railway HPQC is designed at minimum operation voltage under fixed rated load, it can only provide satisfactory at rated load condition and has minimum compensation capability;

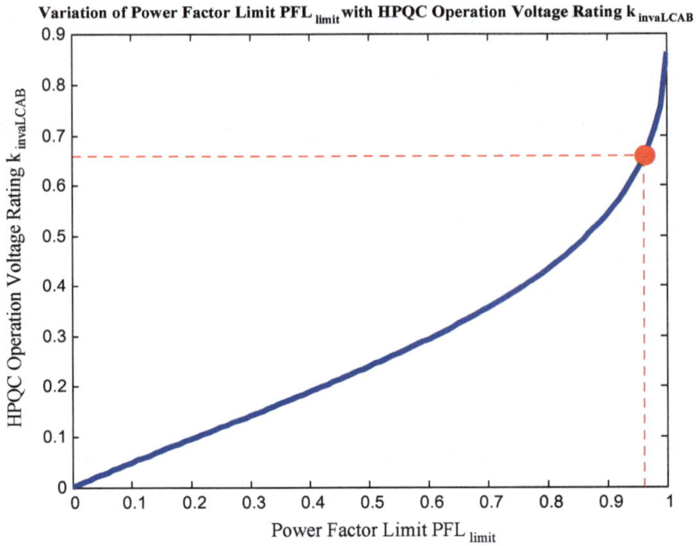

Fig. 4.8 Matlab plot showing the variation of power factor limit PF_{L_limit} with Railway HPQC operation voltage rating k_{invaLC_AB}

- When load varies in practical conditions, the HPQC operation voltage may need to be increased, the relationship between HPQC operation voltage and load condition variation range can then be determined by (4.15);

The expressions in (4.14) can be further investigated and summarized using Fig. 4.6, which is constructed based on rated load power factor value of 0.85. Referring to (4.4) and (4.15), the Railway HPQC compensation range r_B–r_A is a function of HPQC operation voltage rating k_{invaLC_AB} and load power factor PF_L.

The following can be observed from the diagram:

- The value of the compensation range r_B–r_A increases with increase in Railway HPQC operation voltage;
- When load power factor increases, the compensation range decreases;
- There is a limit for load power factor limit, which is investigated in (4.20) and Fig. 4.8, the limit is dependent on k_{invaLC_AB} only;
- The load power factor limit increases with increase in HPQC operation voltage.

4.3.3 Comprehensive Design Procedure for Railway HPQC with Enhanced Compensation Capability

The comprehensive design procedure for Railway HPQC with enhanced compensation capability is as follows:

1. Obtain and compute the value of locomotive PCC voltage V_{ac}, rated compensation current I_{ca_rated}, and V_{ac} phase compensation angle θ_{ca};
2. Determine the value of V_{ac} phase coupled impedance ratio m_{LCa} according to (3.11);
3. Select the value of V_{ac} phase coupled impedance X_{LCa} according to (4.21);

$$|X_{LCa}| = m_{LCa} \cdot \frac{V_{ac}}{I_{ca\_rated}} = \frac{V_{ac} \sin\theta_{ca}}{I_{ca\_rated}} \tag{4.21}$$

4. For harmonic compensation concern, the coupled inductance ratio k_L may be computed based on (3.45);
5. Determine the coupled inductance L_a and capacitance C_a according to the expression in (4.22) and (4.23);

$$L_a = \frac{k_L |X_{LCa}|}{\omega} = \left(\frac{\sum_{h=2}^{\infty} \left(r_{h\_h}^2 \right) \left[\frac{1}{h^2} (h^2 - 1) \right]}{\sum_{h=2}^{\infty} \left(r_{h\_h}^2 \right) \left[\frac{1}{h^2} (h^2 - 1)^2 \right]} \right) \cdot \left(\frac{|X_{LCa}|}{\omega} \right) \tag{4.22}$$

$$C_a = \left(-1 - \frac{\sum_{h=2}^{\infty} \left(r_{h\_h}^2 \right) \left[\frac{1}{h^2} (h^2 - 1) \right]}{\sum_{h=2}^{\infty} \left(r_{h\_h}^2 \right) \left[\frac{1}{h^2} (h^2 - 1)^2 \right]} \right) \cdot \left(\frac{1}{\omega \cdot |X_{LCa}|} \right) \tag{4.23}$$

6. Determine the V_{bc} phase coupled impedance and inductance by (4.29) and (4.30);

$$|X_{LCb}| \geq \frac{V_{bc} \sin\theta_{cb} - \sqrt{V_{invaLC\_min}^2 - V_{bc}^2 \cos^2\theta_{cb}}}{I_{cb}} \tag{4.24}$$

$$L_b = \frac{|X_{LCb}|}{\omega} \tag{4.25}$$

7. Determine the load capacity range in per unit value, and define the lower one as r_A and upper one as r_B;
8. Compute the active and reactive variation ratio h_c and h_s by (4.26);
9. Compute and determine the operation voltage rating of Railway HPQC, k_{invaLC}, according to (4.27);

$$\begin{aligned} h_c &= \frac{\cos\theta_{ca}}{\cos\theta_{ca\_rated}} \\ h_s &= \frac{\sin\theta_{ca}}{\sin\theta_{ca\_rated}} \end{aligned} \tag{4.26}$$

$$k_{\mathrm{invaLC\_AB}} \geq \sqrt{\left(1 - r \cdot h_s \cdot (\sin \theta_{\mathrm{ca\_rated}})^2\right)^2 + (r \cdot h_c \cdot \sin \theta_{\mathrm{ca\_rated}} \cdot \cos \theta_{\mathrm{ca\_rated}})^2} \tag{4.27}$$

10. Compute the Railway HPQC dc Link Voltage according to (4.28)

$$V_{\mathrm{dc}} = \sqrt{2\left(1 - r \cdot h_s \cdot (\sin \theta_{\mathrm{ca\_rated}})^2\right)^2 + 2(r \cdot h_c \cdot \sin \theta_{\mathrm{ca\_rated}} \cdot \cos \theta_{\mathrm{ca\_rated}})^2} \tag{4.28}$$

4.3.4 Simulation Study

The parameter settings in the simulation are chosen based on common practical traction power supply system. The three-phase power grid is of 110 kV, 50 Hz and is transformed into two single-phase outputs via V/V substation transformer. The V/V transformer is composed of two single-phase transformers (31.5 MVA 110 kV/27.5 kV, 31.5 MVA 110 kV/13.75 kV), with V/V connections. One phase of substation output, V_{ac} phase, is connected to locomotive loadings; whereas another phase, V_{bc} phase, is unloaded. The power quality conditioner, Railway HPQC, is then connected across the V_{ac} and V_{bc} phase in order to provide power quality compensation from secondary side to the three-phase primary source grid. Notice that the locomotive voltage is 27.5 kV, which is a bit higher than 25 kV, in order to compromise the voltage drop caused by inductive traction load.

The electronic switches used in the back-to-back converter of Railway HPQC are insulated-gate bipolar transistors (IGBT) for its high power application. The computation of required compensation power is accomplished according to (1) based on instantaneous *pq* theory. The compensation current reference is then obtained by performing inverse transform. The compensation current reference is finally compared with the actual compensation current to generate pulse width modulation (PWM) signals for IGBT switches in Railway HPQC using linear-operated hysteresis PWM method in [30].

In the simulation, the HPQC control algorithm is completed according to (1), and is designed based on minimum operation voltage requirement at fixed rated load of power factor 0.85 and capacity 31.5 MVA (denoted as 1 p.u.). The traction load is then varied from 0.1 to 2 p.u. (0.1 p.u. step size), each with variation of load power factor 0–1 (0.1 step size).

The system parameters used in the simulation are shown in Table 4.1. The V_{ac} phase coupled impedance is calculated using (6) and the coupled LC branch is designed at the fifth load harmonics.

Similar to the condition in the analysis, the value of Railway HPQC operation voltage rating k_{invaLC_AB} used is 0.66. The corresponding dc link voltage is then 25.7 kV. The source power factor and current unbalance are being monitored. The simulated three-phase source power factor and current unbalance (%) under different load conditions obtained are shown in Figs. 4.9 and 4.10.

With satisfactory HPQC compensation, the source power factor is unity and the current unbalance is 0%. It can be observed from Figs. 4.9 and 4.10 that the Railway HPQC can provide satisfactory compensation performance within a range of loadings.

As analyzed, the boundary of the range of loadings which Railway HPQC can provide good compensation performance is defined by (4.15). The corresponding values of r_A, r_M, and r_B calculated with $k_{invaLC\_AB} = 0.66$ under different load power factor PF_L are shown in Table 4.2. The simulated values of r_A and r_B are also shown for comparison. It can be observed that the simulated values are more or less the same as calculated ones. When $PF_L < PF_{L\_rated}$, the value of r_M is greater than 1.0 p.u.; on the other hand, when $PF_L > PF_{L\_rated}$, the value of r_M is smaller than 1.0 p.u.

The limit of load power factor in which Railway HPQC can provide good compensation could also be investigated using (4.20). With $k_{invaLC\_AB} = 0.66$, the value of power factor limit PF_{L_limit} is 0.96.

In the simulation, when the load power factor is 1.0 (exceeding PF_{L_limit}), the system source power factor and current unbalance is not satisfied.

For reference and comparison, the simulations are repeated using (a) conventional RPC with the same dc link voltage at $k_{invaL} = 0.66$, $V_{dc} = 25.7$ kV; (b) Railway HPQC with minimum dc link voltage at rated design, $k_{invaLC} = 0.48$,

Table 4.1 Boundary of calculated and simulated loading conditions which Railway HPQC can provide satisfactory compensation performance with $k_{invaLC\_AB} = 0.66$, $V_{dc} = 25.7$ kV

PF_L	$h_s = r_M$	Calculated r_A	Calculated r_B	Simulated r_A	Simulated r_B
0.1	1.1401	0.3888	1.8914	0.4	1.8
0.2	1.1362	0.3908	1.8816	0.4	1.8
0.3	1.1298	0.3942	1.8653	0.4	1.8
0.4	1.1206	0.3992	1.8420	0.4	1.8
0.5	1.1080	0.4063	1.8097	0.4	1.8
0.6	1.0908	0.4166	1.7650	0.4	1.7
0.7	1.0663	0.4326	1.7001	0.4	1.6
0.8	1.0285	0.4607	1.5963	0.5	1.5
0.9	0.9584	0.5304	1.3865	0.6	1.4
1	0.5708	NA[a]	NA[a]	NA[a]	NA[a]

[a] *NA* Not Available

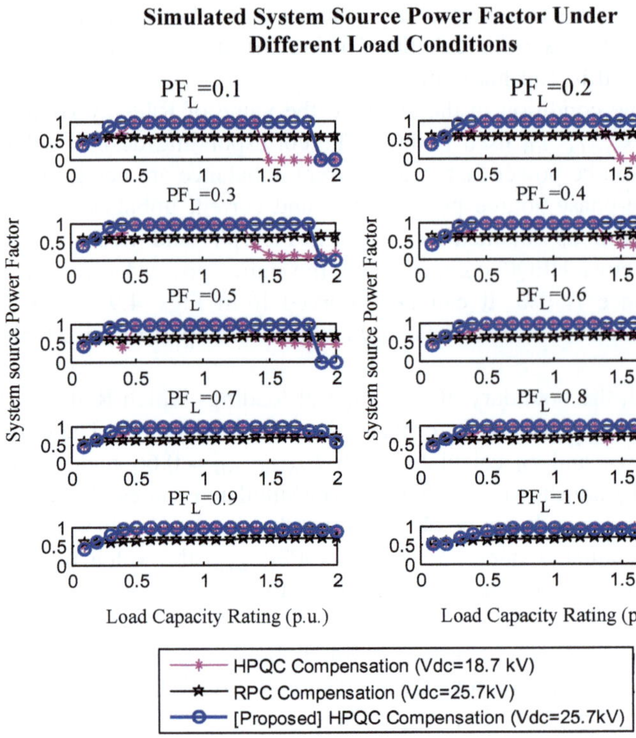

Fig. 4.9 Simulated system source power factor under different loading conditions, obtained from the co-phase traction power with RPC and Railway HPQC under various operation voltage

V_{dc} = 18.7 kV [calculated using (8)]. The results are also presented in Figs. 4.9 and 4.10. It can be observed that consistent with the theoretical study, with conventional RPC operating at the same dc link voltage, the system performance is far from satisfactory at any conditions. It could also be seen from the figures that the compensation capability is limited when using Railway HPQC with operation voltage at its minimum value at rated load. The analysis of Railway HPQC compensation capability is, therefore, verified via PSCAD simulations. They also show the significance of increasing Railway HPQC operation voltage using proposed analysis in enhancing the compensation capability.

4.3.5 Experimental Results

In order to verify the increase in Railway HPQC compensation capability, experimental results are obtained from the hardware prototype. It is assumed that the traction load capacity changes from 0.2 to 1.2 p.u. Detailed design procedures can

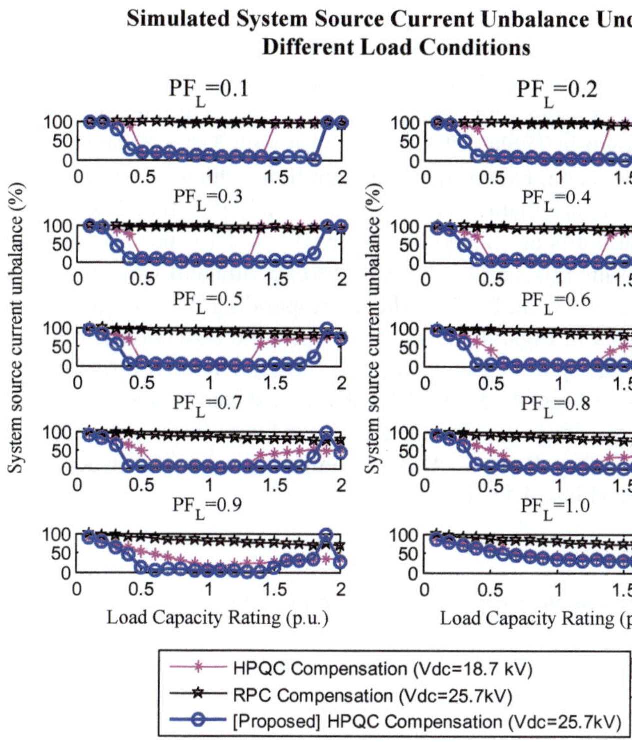

Fig. 4.10 Simulated system source current unbalance (%) under different loading conditions, obtained from the co-phase traction power with RPC and Railway HPQC under various operation voltage

Table 4.2 Experimental parameter design of novel Railway HPQC under traditional harmonic filter design for verification

No.	Parameters	Symbol	Value
1.	V_{ac} phase coupled impedance ratio	m_{LCa}	0.876
2.	V_{ac} phase coupled inductance ratio	k_L	0.10
3.	V_{ac} phase coupled inductance	L_a	4.9 mH
4.	V_{ac} phase coupled capacitance	C_a	170 μF
5.	V_{bc} phase coupled inductance	L_b	4 mH
6.	dc link capacitance	C_{dc}	10,000 μF
7.	Railway HPQC operation voltage rating	k_{invaLC}	0.66
8.	dc link operation voltage	V_{dc}	47 V

be referred to Sect. 4.3.3. First, the coupled impedance X_{LCa}, coupled inductance ratio k_L, coupled inductance L_a, and coupled capacitance C_a are all selected based on the discussion in Chap. 3. Refer to (3.51)–(3.53) for details. However, in order to enhance the Railway HPQC compensation capability, the operation voltage is enhanced to a higher level. The required Railway HPQC operation voltage rating may then be computed based on (4.14). In this investigation, it is assumed that the load power factor does not change, $h_c = h_s = 1$. The detailed system design parameter is shown in Table 4.2. Comparing it with Table 3.10, it can be seen that everything is the same, except that the operation voltage is enhanced.

Four conditions are mainly investigated, namely 0.2, 0.5, 1.0, and 1.7 p.u. The middle two conditions are inside the range (0.5 and 1.0 p.u.), while the other two lies outside the range (0.2 and 1.7 p.u.). The experimental waveforms are shown in Figs. 4.11, 4.12, 4.13, and 4.14. The corresponding power quality statistics are shown as graphics in Figs. 4.15 and 4.16. It can be observed from the waveforms that for load capacity of 0.5 and 1.0 p.u., the system waveforms are satisfactory. The data points in Figs. 4.15 and 4.16 also reveal that under these conditions, the three-phase source power factor is near unity and with low current unbalance. These verify the satisfactory Railway HPQC compensation performance and capability within the range (Figs. 4.15 and 4.16).

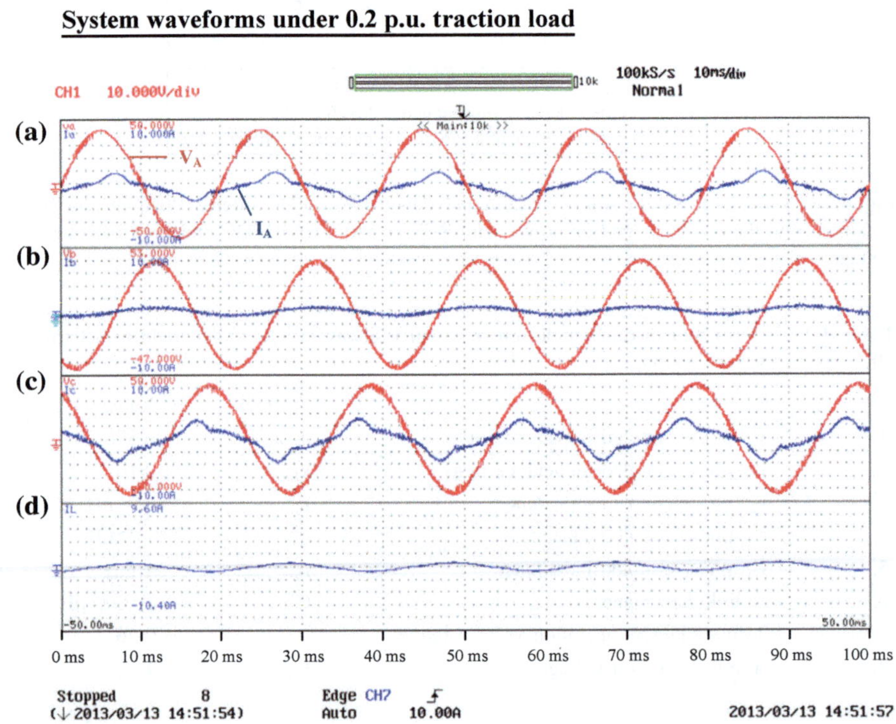

Fig. 4.11 Experimental system waveforms obtained from the co-phase traction power hardware prototype under 0.2 p.u. of rated load, source voltage, and current of **a** phase A; **b** phase B; **c** phase C; **d** load current

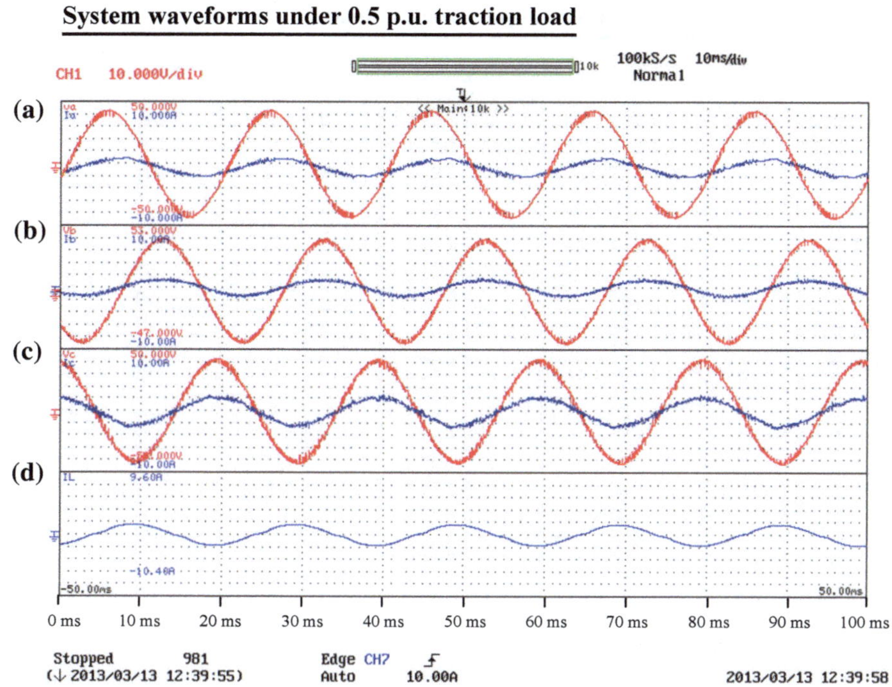

Fig. 4.12 Experimental system waveforms obtained from the co-phase traction power hardware prototype under 0.5 p.u. of rated load, source voltage, and current of **a** phase A; **b** phase B; **c** phase C; **d** load current

On the other hand, for traction load capacity of 0.2 and 1.7 p.u., the system waveforms get distorted, and the phase angle between phase voltage and current also gets larger. It can also be seen from the data point in Figs. 4.20 and 4.21 that at load capacity of 0.2 and 1.7 p.u., the system source power factor and current unbalance gets worse. This indicates the unsatisfactory HPQC compensation performance at these load capacities (0.2, 1.7 p.u.).

Referring to the simulation results presented in Table 4.1; Figs. 4.9 and 4.10, it can be observed that for load power factor of 0.8 and 0.9, with $k_{\mathrm{invaLC}} = 0.66$, the boundary of the range of loading capacity that satisfactory compensation can be provided are $r = [0.5, 1.5]$ and $[0.6, 1.4]$, respectively.

4.4 Impedance-Mapping Technique According to Load Variation Range (for Reduced Operation Voltage)

In order to provide satisfactory compensation performance when load varies from rated designed value, the coupled impedance may be selected such that it will match the load variation range. Referring to the expression of Railway HPQC operation

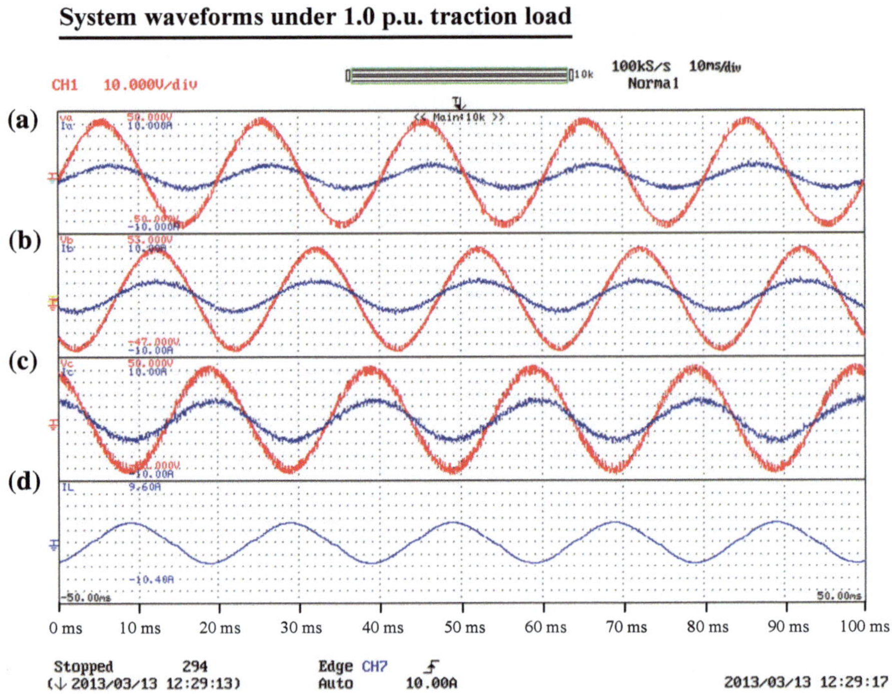

Fig. 4.13 Experimental system waveforms obtained from the co-phase traction power hardware prototype under 1.0 p.u. of rated load, source voltage, and current of **a** phase A; **b** phase B; **c** phase C; **d** load current

voltage determination in (4.7), by substituting the definition of X_{LCa} in (2.42) into (4.7), the expression can be revised as in (4.29). It can be observed from the expression that besides increasing the operation voltage rating k_{invaLC}, the coupled impedance, which is reflected by the value of coupled impedance ratio m_{LCa}, can also be modified to provide good compensation performance.

$$
\begin{aligned}
k_{\text{invaLC}} &= \frac{\left|\vec{V}_{\text{invaLC}}\right|}{V_{\text{ac}}} \\
&= \sqrt{\left(1 - r \cdot h_s \cdot \sin\theta_{\text{ca\_rated}} \cdot m_{\text{LCa}}\right)^2 + \left(r \cdot h_c \cdot \cos\theta_{\text{ca\_rated}} \cdot m_{\text{LCa}}\right)^2}
\end{aligned}
\tag{4.29}
$$

4.4.1 Concept of Mapping Railway HPQC Coupled Impedance with Load Variation Range

With (4.29), it can be observed that besides selecting the value of m_{LCa} for minimum Railway HPQC operation voltage at rated designed load capacity and condition, the value of m_{LCa} may be chosen according to the load variation range

System waveforms under 1.7 p.u. traction load

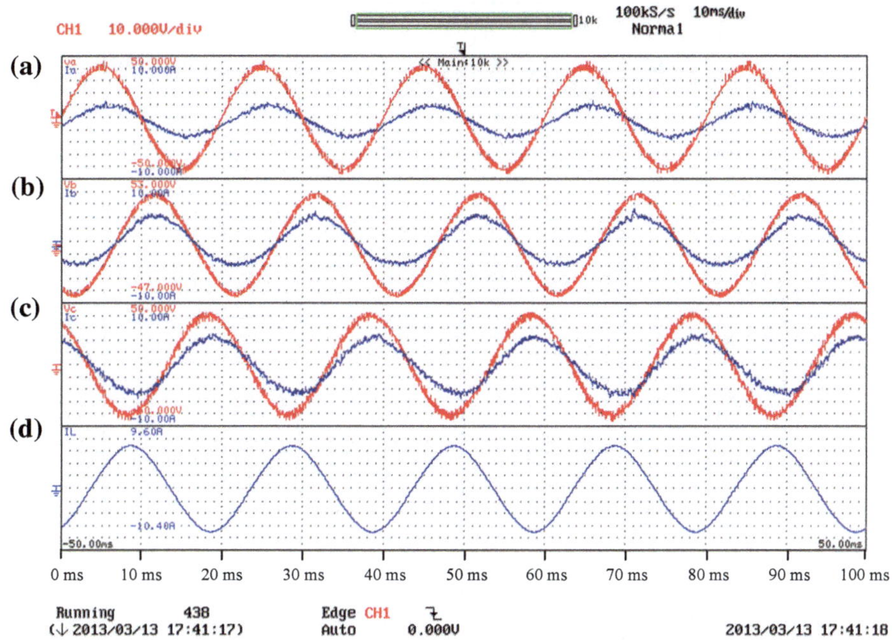

Fig. 4.14 Experimental system waveforms obtained from the co-phase traction power hardware prototype under 1.7 p.u. of rated load, source voltage, and current of **a** phase A; **b** phase B; **c** phase C; **d** load current

parameters r and h_s. By further manipulation of (4.29), the relationship between the value of m_{LCa} and other parameters can be obtained, as in (4.30).

$$r^2(m_{\mathrm{LCa}})^2 - m_{\mathrm{LCa}} \cdot (2 \cdot r \cdot h_s \cdot \sin\theta_{\mathrm{ca\_rated}}) + \left(1 - k_{\mathrm{invaLC}}^2\right) = 0 \qquad (4.30)$$

The boundary conditions (point A and B in Fig. 4.7) expressions are shown in (4.31). In Railway HPQC, the value of coupled impedance ratio $m_{\mathrm{LCa\_AB}}$ and operation voltage rating $k_{\mathrm{invaLC\_AB}}$ should be the same at boundary conditions.

$$\begin{cases} (r_A)^2(m_{\mathrm{LCa\_AB}})^2 - m_{\mathrm{LCa\_AB}} \cdot (2 \cdot r_A \cdot h_{\mathrm{sA}} \cdot \sin\theta_{\mathrm{ca\_rated}}) + \left(1 - k_{\mathrm{invaLC\_AB}}^2\right) = 0 \\ (r_B)^2(m_{\mathrm{LCa\_AB}})^2 - m_{\mathrm{LCa\_AB}} \cdot (2 \cdot r_B \cdot h_{\mathrm{sB}} \cdot \sin\theta_{\mathrm{ca\_rated}}) + \left(1 - k_{\mathrm{invaLC\_AB}}^2\right) = 0 \end{cases} \qquad (4.31)$$

By solving the value of m_{LCa} and $k_{\mathrm{invaLC\_AB}}$ in (4.31), the determination expressions can be obtained as in (4.32).

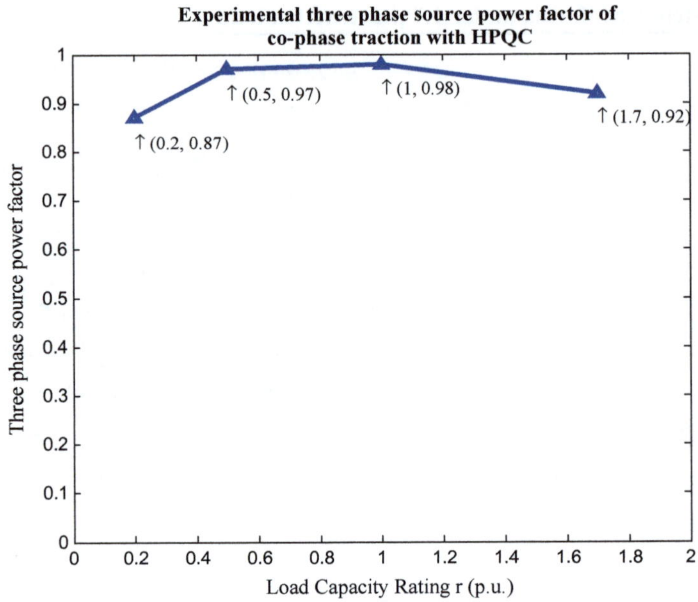

Fig. 4.15 Experimental data of system source power factor of co-phase traction with novel Railway HPQC under 0.2, 0.5, 1.0, and 1.7 p.u

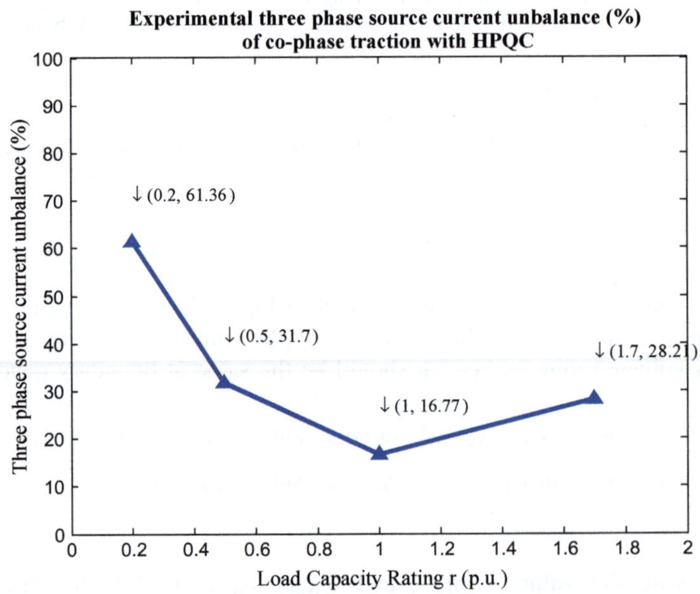

Fig. 4.16 Experimental data of system source current unbalance (%) of co-phase traction with novel Railway HPQC under 0.2, 0.5, 1.0, and 1.7 p.u

$$\begin{cases} m_{\mathrm{LCa\_AB}} = \frac{2 \cdot \sin\theta_{\mathrm{ca\_rated}} \cdot (r_A h_{\mathrm{sA}} - r_B h_{\mathrm{sB}})}{(r_A + r_B)(r_A - r_B)} = g \cdot \sin\theta_{\mathrm{ca\_rated}} = g \cdot m_{\mathrm{LCa\_rated}} \\ k_{\mathrm{invaLC\_AB}} = \sqrt{(r_A)^2 (m_{\mathrm{LCa\_AB}})^2 - m_{\mathrm{LCa\_AB}} \cdot (2 \cdot r_A \cdot h_{sA} \cdot \sin\theta_{\mathrm{ca\_rated}}) + 1} \\ = \sqrt{(r_B)^2 (m_{\mathrm{LCa\_AB}})^2 - m_{\mathrm{LCa\_AB}} \cdot (2 \cdot r_B \cdot h_{\mathrm{sB}} \cdot \sin\theta_{\mathrm{ca\_rated}}) + 1} \end{cases} \tag{4.32}$$

4.4.2 Reduction in Coupled Capacitance

Based on the expression in (4.32) and the definition of $m_{\mathrm{LCa\_rated}}$ in (2.42) and (3.10), it could be observed that using impedance-mapping technique, the coupled impedance X_{LCa} is modified by a factor of g, which is determined by (4.33).

$$g = \frac{2 \cdot (r_A h_{\mathrm{sA}} - r_B h_{\mathrm{sB}})}{(r_A + r_B)(r_A - r_B)} \tag{4.33}$$

It is obvious that when the coupled impedance X_{LCa} is changed, the coupled capacitance will also be changed. The effect on coupled capacitance using impedance-mapping technique is explored below.

First of all, the coupled impedance is defined as in (4.34).

$$X_{\mathrm{LCa\_AB}} = X_{\mathrm{La}} + X_{\mathrm{Ca}} = (k_L + k_C)|X_{\mathrm{LCa\_AB}}| = (k_L + k_C) \cdot g \cdot |X_{\mathrm{LCa\_rated}}| \tag{4.34}$$

The value of k_L and k_C is determined according to load harmonic conditions and is assumed to be constant in this analysis. Therefore, the coupled capacitance impedance is defined as in (4.35).

$$|X_{\mathrm{Ca}}| = k_C \cdot g \cdot |X_{\mathrm{LCa\_rated}}| \tag{4.35}$$

The coupled capacitance C_a in impedance-mapping method may then be expressed as (4.36). It can be observed that using impedance-mapping technique, the coupled capacitance is changed to $1/g$ times of rated value.

$$\begin{aligned} C_a &= \frac{1}{\omega \cdot |X_{\mathrm{Ca}}|} = \frac{1}{\omega \cdot k_C \cdot g \cdot |X_{\mathrm{LCa\_rated}}|} \\ &= \frac{1}{\omega \cdot g \cdot |X_{\mathrm{Ca\_rated}}|} = \frac{\omega \cdot C_{\mathrm{a\_rated}}}{\omega \cdot g} = \frac{1}{g} \cdot C_{\mathrm{a\_rated}} \end{aligned} \tag{4.36}$$

Therefore, when the value of g is larger than 1, the coupled capacitance is reduced; on the other hand, when the value of g is smaller than 1, the coupled capacitance will be increased. Reduction of coupled capacitance can help to reduce cost. Thus, the impedance-mapping technique is preferred only when $g > 1$.

Further analysis is performed below. The expression for determining value of g is shown in (4.33). For g to be larger than 1, the expression in (4.37) must be satisfied.

$$2 \cdot (r_A h_{sA} - r_B h_{sB}) > (r_A + r_B)(r_A - r_B) \tag{4.37}$$

4.4.3 Reduction in Railway HPQC Operation Voltage Rating

It can also be inferred from (4.32) that when the value of m_{LCa} is changed to m_{LCa_AB}, the Railway HPQC operation voltage rating k_{invaLC} will also be changed. The change in the operation voltage will be explored in the following.

The ratio between the Railway HPQC operation voltage using impedance-mapping technique in (4.32) and that using design based on minimum operation requirement in (3.10) is given in (4.38).

$$\frac{k_{invaLC\_AB}}{k_{invaLC\_rated}} = \frac{\sqrt{(r_A)^2(g)^2(m_{LCa\_rated})^2 - g \cdot m_{LCa\_rated} \cdot (2 \cdot r_A \cdot h_{sA} \cdot \sin\theta_{ca\_rated}) + 1}}{\sqrt{(r_A)^2(m_{LCa\_rated})^2 - m_{LCa\_rated} \cdot (2 \cdot r_A \cdot h_{sA} \cdot \sin\theta_{ca\_rated}) + 1}} = \frac{\sqrt{(r_B)^2(g)^2(m_{LCa\_rated})^2 - g \cdot m_{LCa\_rated} \cdot (2 \cdot r_B \cdot h_{sB} \cdot \sin\theta_{ca\_rated}) + 1}}{\sqrt{(r_B)^2(m_{LCa\_rated})^2 - m_{LCa\_rated} \cdot (2 \cdot r_B \cdot h_{sB} \cdot \sin\theta_{ca\_rated}) + 1}} \tag{4.38}$$

By substituting the value of m_{LCa_rated} in (3.10) into (4.38), the expression in (4.38) can be revised as (4.39).

$$\frac{k_{invaLC\_AB}}{k_{invaLC\_rated}} = \frac{\sqrt{(r_A)^2(g)^2(\sin\theta_{ca\_rated})^2 - g \cdot (\sin\theta_{ca\_rated})^2 \cdot (2 \cdot r_A \cdot h_{sA}) + 1}}{\sqrt{(r_A)^2(\sin\theta_{ca\_rated})^2 - (\sin\theta_{ca\_rated})^2 \cdot (2 \cdot r_A \cdot h_{sA} \cdot) + 1}} = \frac{\sqrt{(r_B)^2(g)^2(\sin\theta_{ca\_rated})^2 - g \cdot (\sin\theta_{ca\_rated})^2 \cdot (2 \cdot r_B \cdot h_{sB}) + 1}}{\sqrt{(r_B)^2(\sin\theta_{ca\_rated})^2 - (\sin\theta_{ca\_rated})^2 \cdot (2 \cdot r_B \cdot h_{sB}) + 1}} \tag{4.39}$$

For analysis simplicity, it is first assumed that the locomotive load power factor is nearly constant such that $h_{sA} = h_{sB} = 1$. Assuming common locomotive load power 0.85 as the rated designed load condition, a Matlab plot is constructed based on (4.39) and is shown in Fig. 4.17.

The x axis is the load capacity ratio and goes from zero to rated value since it represents the lower boundary capacity r_A, which is usually lower than rated value at normal case.

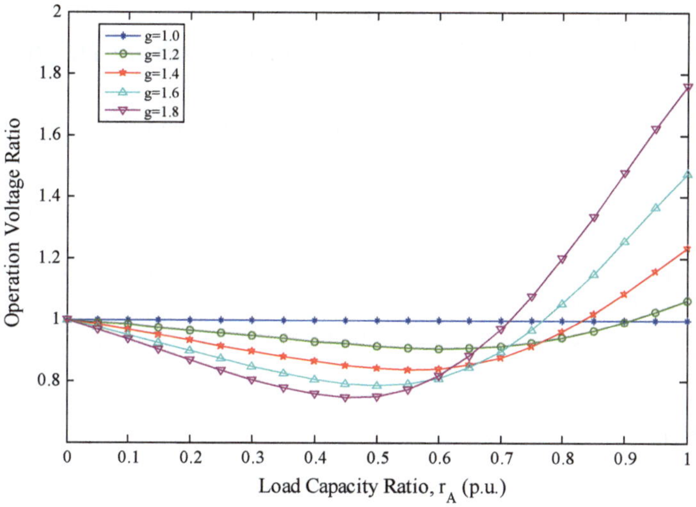

Fig. 4.17 Variation of ratio between Railway HPQC operation voltage using impedance-mapping technique and that using design based on minimum operation requirement with parameters *g* and *r*

The followings can be observed from the figure:

- Reduction of operation voltage rating using impedance-mapping technique can be achieved most of the time when $g > 1$ (when reduction in capacitance is achieved);
- The relationship between load capacity ratio and operation voltage rating using impedance-mapping technique is not linear;

For the Railway HPQC operation voltage rating using impedance-mapping technique to be lower than that using design based on minimum operation voltage at rated load condition, the value of g in (4.38) must be lower than 1.0 such that the expression in (4.40) is satisfied.

$$k_{\text{invaLC\_AB}} < k_{\text{invaLC\_rated}} \tag{4.40}$$

By further manipulation of (4.40), the expression in (4.41) can be obtained. Reduction in operation voltage rating can be achieved using impedance-mapping technique when the condition in (4.41) is satisfied

$$\begin{aligned} r_A\left(\frac{2(r_A h_{\text{sA}} - r_B h_{\text{sB}})}{(r_A + r_B)(r_A - r_{\text{B}})} + 1\right) &< h_{\text{sA}}, \text{ or} \\ r_{\text{B}}\left(\frac{2(r_A h_{\text{sA}} - r_B h_{\text{sB}})}{(r_A + r_B)(r_A - r_{\text{B}})} + 1\right) &< h_{\text{sB}} \end{aligned} \tag{4.41}$$

4.4.4 Comprehensive Design Procedures for Impedance-Mapping Co-phase Railway HPQC

The comprehensive design procedure for Railway HPQC based on impedance-mapping technique is shown below:

1. Obtain and compute the value of locomotive PCC voltage V_{ac}, rated compensation current I_{ca_rated}, and V_{ac} phase compensation angle θ_{ca};
2. Determine the load capacity range in per unit value, and define the lower one as r_A and upper one as r_B;
3. Determine the value of V_{ac} phase coupled impedance ratio m_{LCa_AB} according to (4.42);

$$m_{LCa\_AB} = g \cdot m_{LCa\_rated} = \frac{2 \cdot (r_A h_{sA} - r_B h_{sB})}{(r_A + r_B)(r_A - r_B)} \cdot \sin \theta_{ca\_rated} \tag{4.42}$$

4. Select the value of V_{ac} phase coupled impedance X_{LCa} according to (4.43);

$$|X_{LCa}| = m_{LCa\_AB} \cdot \frac{V_{ac}}{I_{ca\_rated}} = \frac{2 \cdot (r_A h_{sA} - r_B h_{sB})}{(r_A + r_B)(r_A - r_B)} \cdot \sin \theta_{ca\_rated} \cdot \frac{V_{ac} \sin \theta_{ca\_rated}}{I_{ca\_rated}} \tag{4.43}$$

5. For harmonic compensation concern, the coupled inductance ratio k_L may be computed based on (4.45);
6. Determine the coupled inductance L_a and capacitance C_a according to the expression in (4.44) and (4.45);

$$L_a = \frac{k_L |X_{LCa}|}{\omega} = \left(\frac{\sum_{h=2}^{\infty} \left(r_{h\_h}^2 \right) \left[\frac{1}{h^2} (h^2 - 1) \right]}{\sum_{h=2}^{\infty} \left(r_{h\_h}^2 \right) \left[\frac{1}{h^2} (h^2 - 1)^2 \right]} \right) \cdot \left(\frac{|X_{LCa}|}{\omega} \right) \tag{4.44}$$

$$C_a = \left(-1 - \frac{\sum_{h=2}^{\infty} \left(r_{h\_h}^2 \right) \left[\frac{1}{h^2} (h^2 - 1) \right]}{\sum_{h=2}^{\infty} \left(r_{h\_h}^2 \right) \left[\frac{1}{h^2} (h^2 - 1)^2 \right]} \right) \cdot \left(\frac{1}{\omega \cdot |X_{LCa}|} \right) \tag{4.45}$$

7. Determine the V_{bc} phase coupled impedance and inductance by (4.46) and (4.47);

$$|X_{LCb}| \geq \frac{V_{bc}\sin\theta_{cb} - \sqrt{V_{invaLC\_min}^2 - V_{bc}^2\cos^2\theta_{cb}}}{I_{cb}} \tag{4.46}$$

$$L_b = \frac{|X_{LCb}|}{\omega} \tag{4.47}$$

8. Compute the active and reactive variation ratio h_c and h_s by (4.26);
9. Compute and determine the operation voltage rating of Railway HPQC, k_{invaLC}, according to (4.49);

$$\begin{aligned} h_c &= \frac{\cos\theta_{ca}}{\cos\theta_{ca\_rated}} \\ h_s &= \frac{\sin\theta_{ca}}{\sin\theta_{ca\_rated}} \end{aligned} \tag{4.48}$$

$$\begin{aligned} k_{invaLC\_AB} &= \sqrt{(r_A)^2(m_{LCa\_AB})^2 - m_{LCa\_AB}\cdot(2\cdot r_A\cdot h_{sA}\cdot\sin\theta_{ca\_rated}) + 1} \\ &= \sqrt{(r_B)^2(m_{LCa\_AB})^2 - m_{LCa\_AB}\cdot(2\cdot r_B\cdot h_{sB}\cdot\sin\theta_{ca\_rated}) + 1} \end{aligned} \tag{4.49}$$

$$\begin{aligned} k_{invaLC\_AB} &= \sqrt{(r_A)^2(m_{LCa\_AB})^2 - m_{LCa\_AB}\cdot(2\cdot r_A\cdot h_{sA}\cdot\sin\theta_{ca\_rated}) + 1} \\ &= \sqrt{(r_B)^2(m_{LCa\_AB})^2 - m_{LCa\_AB}\cdot(2\cdot r_B\cdot h_{sB}\cdot\sin\theta_{ca\_rated}) + 1} \end{aligned} \tag{4.50}$$

10. Compute the Railway HPQC dc Link Voltage according to (4.50);

$$\begin{aligned} V_{dc} &= \sqrt{2(r_A)^2(m_{LCa\_AB})^2 - 2m_{LCa\_AB}\cdot(2\cdot r_A\cdot h_{sA}\cdot\sin\theta_{ca\_rated}) + 2} \\ &= \sqrt{2(r_B)^2(m_{LCa\_AB})^2 - 2m_{LCa\_AB}\cdot(2\cdot r_B\cdot h_{sB}\cdot\sin\theta_{ca\_rated}) + 2} \end{aligned} \tag{4.50}$$

4.4.5 Simulation and Case Study

Simulations verifications are done using PSCAD to show the reduction of operation voltage and coupled capacitance using impedance-mapping Railway HPQC for providing compensation in co-phase traction power within the designed load variation range. Shown in Fig. 3.15 is the simulation circuit schematic of the co-phase traction power supply under investigation. Practical conditions are taken into consideration to simulate real implementation. The circuit structure resembles that of China WuQing Substation, except that the co-phase technique is adopted. The substation transformer is in V/V connection and is composed of two single-phase transformers, each of 31.5 MVA. As described previously, in co-phase traction power, the traction load is connected across V_{ac} phase only, and the power quality conditioner, Railway HPQC, is connected across V_{ac} and V_{bc} phase to provide compensation.

The simulation is performed with various loading capacities. With rated capacity at 31.5 MVA, the traction load capacity ranges from 6.5 to 36.5 MVA (0.2–1.2 p.u.), which is 0.21–1.16 in per unit value. It is further assumed that similar to other traction loadings, the load power factor is around 0.85. In the verification, the traction load is simulated using rectifier RLC load, with load capacity range of 0–2.0 p.u. It is further assumed that the load capacity varies with constant load power factor such that $h_s = h_c = 1.0$.

The performance of Railway HPQC with two conditions, namely, (1) Railway HPQC with enhanced operation voltage based on coupled impedance designed based on minimum operation voltage requirement at rated load condition; and (2) Railway HPQC with impedance-mapping technique for reduction in coupled capacitance and operation voltage are being investigated. The three-phase system source power factor (PF), current unbalance and total harmonics distortions (THD) are monitored according to the IEEE standard. The main goal of Railway HPQC under this simulation verification is to provide satisfactory compensation performance within the designed load variation range (0.2–1.2 p.u.).

Railway HPQC with enhanced operation voltage based on coupled impedance designed based on minimum operation voltage requirement at rated load condition ($V_{dc} = 34.5$ kV)

Simulations are first done with Railway HPQC with enhanced operation voltage with coupled impedance developed based on the requirement of minimum operation voltage at rated load condition. The Vac coupled impedance ratio m_{LCa} and Railway HPQC operation voltage rating is designed according to (3.11), (3.51), (4.14), and (4.28). Details of HPQC design and component parameters can be found in Table 4.3.

Simulation with Railway HPQC using impedance-mapping design based on load variation range ($r_A = 0.2$, $r_B = 1.2$, $h_s = 1$, $h_c = 1$, $V_{dc} = 31.5$ kV).

Next, simulation results are obtained with Railway HPQC using impedance-mapping technique based on load variation range. In contrast with minimum operation voltage HPQC design, the Railway HPQC is designed to provide satisfactory compensation for desired load capacity of 0.2–1.2 p.u. Assuming that the load capacity varies with fixed rated load power factor, the V_{ac} coupled impedance ratio m_{LCa_AB} and Railway HPQC operation voltage rating may be designed according to (4.42) and (4.49) as well as the procedures in Sect. 4.4.4. The Railway HPQC parameters with impedance-mapping design are presented in Table 4.4.

Table 4.3 Railway HPQC design with minimum operation voltage for rated load of WuQing Substation based on minimum operation voltage requirement

Parameters	Descriptions	Value
$I_{\mathrm{ca\_rated}}$	V_{ac} phase compensation current	1000 A
PF_L	Load power factor	0.85
θ_{ca}	V_{ac} phase compensation angle	61.28
g	V_{ac} phase coupled impedance ratio modification factor	1.0
m_{LCa}	V_{ac} coupled inductance ratio	0.87
L_a	V_{ac} coupled inductance	7.46 mH
C_a	V_{ac} coupled capacitance	123 μF
k_{invaL}	Railway HPQC operation voltage rating	0.83
V_{dc}	dc link voltage	34.5 kV

Table 4.4 Detailed parameters of impedance-mapping Railway HPQC design for 0.2–1.2 p.u. traction load capacity variation range

Parameters	Descriptions	Value
r_a	Load capacity ratio	0.2
r_b	Load capacity ratio	1.2
I_{ca}	V_{ac} phase compensation current	1000 A
PF_L	Load power factor	0.85
θ_{ca}	V_{ac} phase compensation angle	61.28
g	V_{ac} phase coupled impedance ratio modification factor	1.44
m_{LCa}	V_{ac} coupled impedance ratio	1.26
L_a	V_{ac} coupled inductance	10.9 mH
C_a	V_{ac} coupled capacitance	84 μF
k_{invaLC}	Railway HPQC operation voltage rating	0.79
V_{dc}	dc link voltage	31.5 kV

The simulated system waveforms are shown in Fig. 4.18. Four conditions are mainly presented, namely, (a) outside designed compensation range near no load conditions (0.1 p.u.), (b) within designed compensation range (0.6 p.u.); (c) within designed compensation range at rated load (1.0 p.u.), (d) outside designed compensation range at overload (1.3 p.u.). More simulation performance of source power factor, current harmonic distortions, and system unbalance under different load capacity are shown as graphics in Figs. 4.19, 4.20, and 4.21.

It can be observed from the figure that for Railway HPQC developed based on minimum operation at rated load, the compensation performance is satisfactory even when outside the compensation range (0.1, 1.3 p.u.), indicating that the compensation capability is more than enough. This causes high excessive operation voltage and coupled capacitance. On the other hand, with the impedance-mapping Railway HPQC, satisfactory compensation performance is obtained within the load range (0.2–1.2 p.u.), but with a 9% reduction in operation voltage. Moreover, there is around

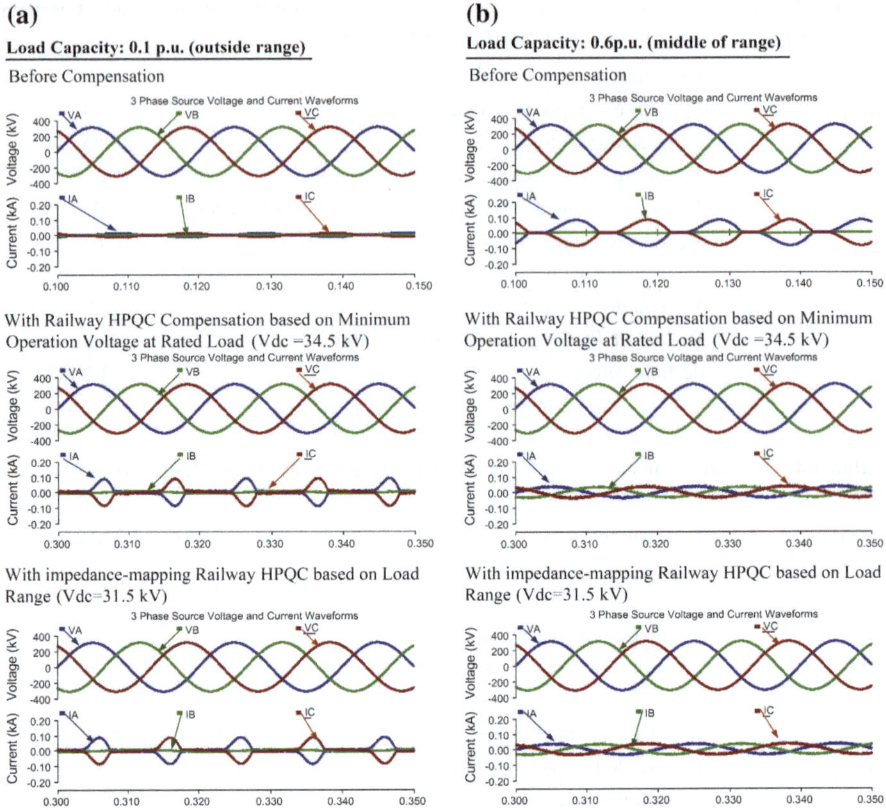

Fig. 4.18 Simulated co-phase traction power system waveforms under different load rating and simulation conditions: **a** 0.1 p.u.; **b** 0.6 p.u.; **c** 1.0 p.u.; and **d** 1.3 p.u

30% reduction in the coupled capacitance. This is consistent with previous analysis and shows the effectiveness of impedance-mapping Railway HPQC in reducing operation voltage and coupled capacitance. Notice that the system performance is more or less the same within the load variation range (0.2–1.2 p.u.).

4.4.6 Experimental Results

Similar to previous analysis, experimental results are obtained to verify the performance of co-phase traction power supply with Railway HPQC using impedance-mapping technique. It is assumed that the desired compensation range is from 0.2 to 1.2 p.u., same as the traction load variation range. It is further assumed that the load power factor is kept constant such that $h_c = h_s = 1$. The system parameter is designed according to the procedures in Sect. 4.4.4 and is shown in

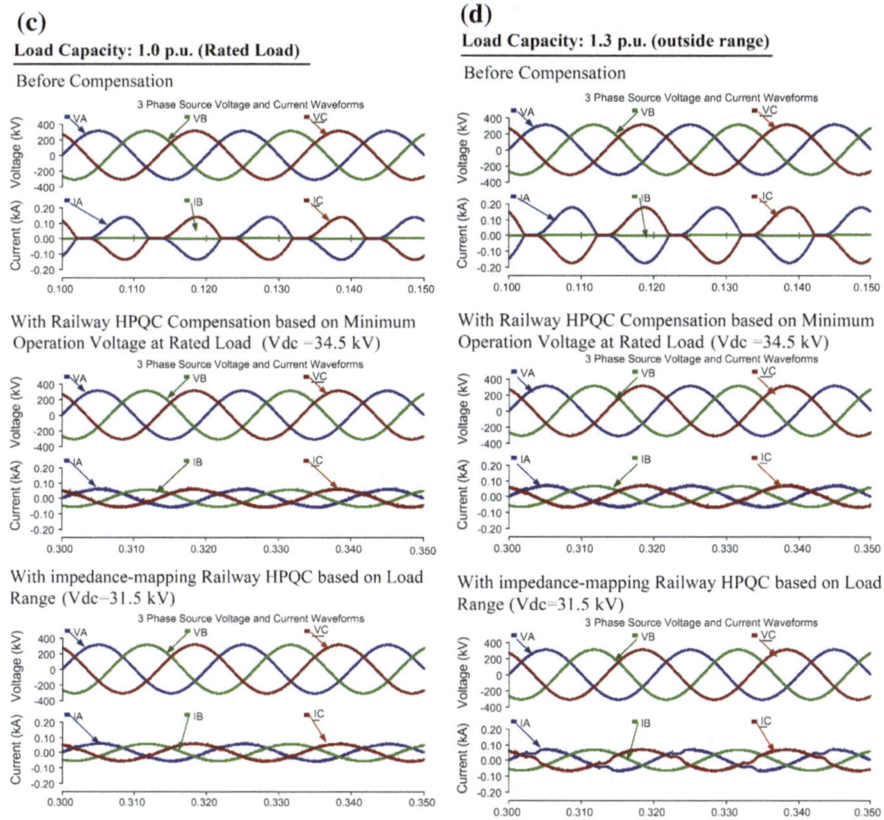

Fig. 4.18 (continued)

Table 4.5. Unlike other designs, the coupled impedance ratio m_{LCa} is determined based on the load variation condition ($r_A = 0.2$, $r_B = 1.2$) according to (4.42). The coupled impedance X_{LCa}, coupled inductance L_a and capacitance C_a may then be computed according to (3.52), (3.53), and (4.43). Finally, the Railway HPQC operation voltage rating may then be computed based on (4.49).

Four load capacity conditions are investigated: 0.2 (boundary of compensation range), 0.6 (inside and middle of compensation range), 1.0 (inside compensation range and rated load), and 1.4 (outside compensation range). The system waveforms obtained are shown in Figs. 4.22, 4.23, 4.24, 4.25, 4.26, 4.27, 4.28, 4.29, 4.30, and 4.31. The data of power factor, current THD, and unbalance is shown in Figs. 4.26, 4.27, and 4.28. It can be verified that using Railway HPQC under impedance-mapping design, the compensation performance is satisfactory within the desired compensation range (0.2–1.2 p.u.). Moreover, compared to traditional method, there is over 10% reduction in operation voltage and 30% reduction in coupled capacitance.

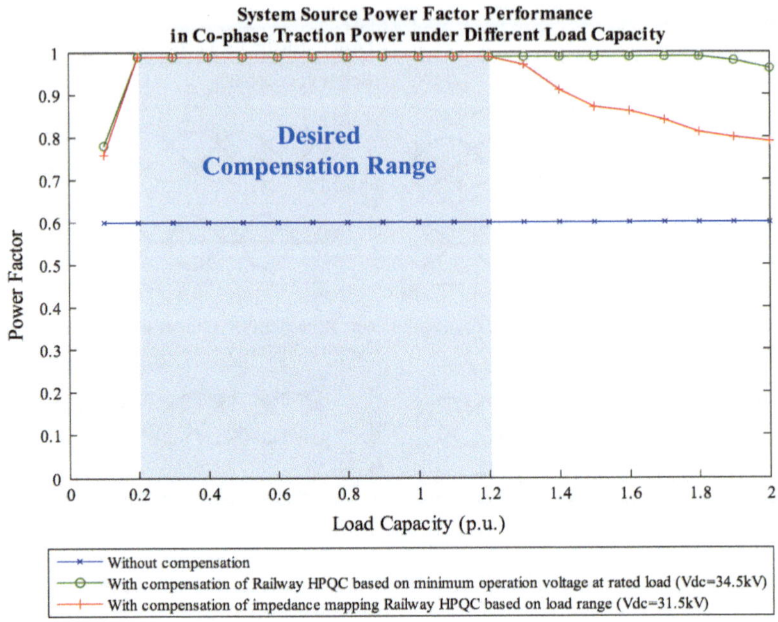

Fig. 4.19 Simulated system source power factor performance in co-phase traction power under different load capacity

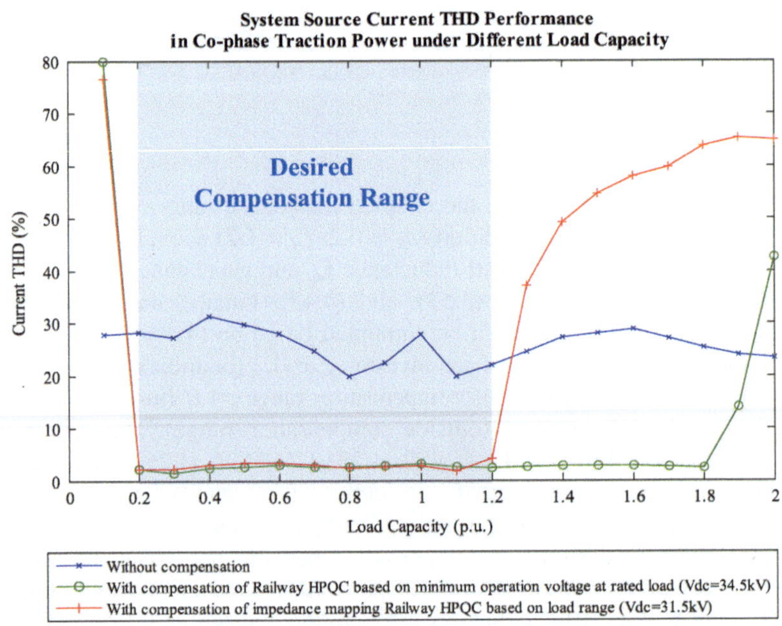

Fig. 4.20 Simulated system source current THD performance in co-phase traction power under different load capacity

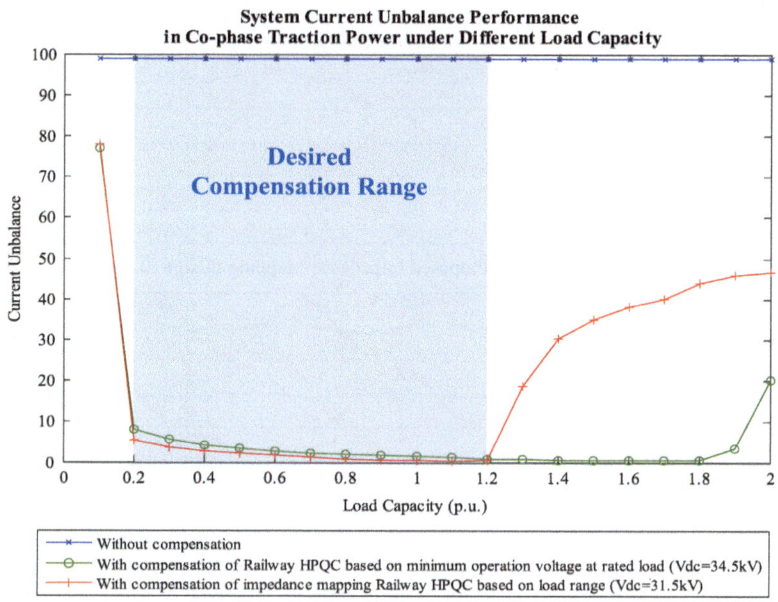

Fig. 4.21 Simulated system source current unbalance performance in co-phase traction power under different load capacity

Table 4.5 Experimental parameter design of Railway HPQC under impedance-mapping design for verification

No.	Parameters	Symbol	Value
1.	V_{ac} phase coupled impedance ratio	m_{LCa}	1.25 (g = 1.43)
2.	V_{ac} phase coupled inductance ratio	k_L	0.10
3.	V_{ac} phase coupled inductance	L_a	7.9 mH
4.	V_{ac} phase coupled capacitance	C_a	122 μF
5.	V_{bc} phase coupled inductance	L_b	4 mH
6.	dc link capacitance	C_{dc}	10,000 μF
7.	Railway HPQC operation voltage rating	k_{invaLC}	0.79
8.	dc link operation voltage	V_{dc}	57 V

4.5 Adaptive dc Link Control Technique for Co-phase Railway HPQC for Load Variations

So far, two methods for Railway HPQC design based on load variations are discussed. The operation voltage rating is enhanced to certain level in the first method of enhancing Railway HPQC operation voltage to increase compensation capability as in (4.14). However, the operation voltage rating design is kept constant based on load variation range. Excessive operation voltage may be used when there is little

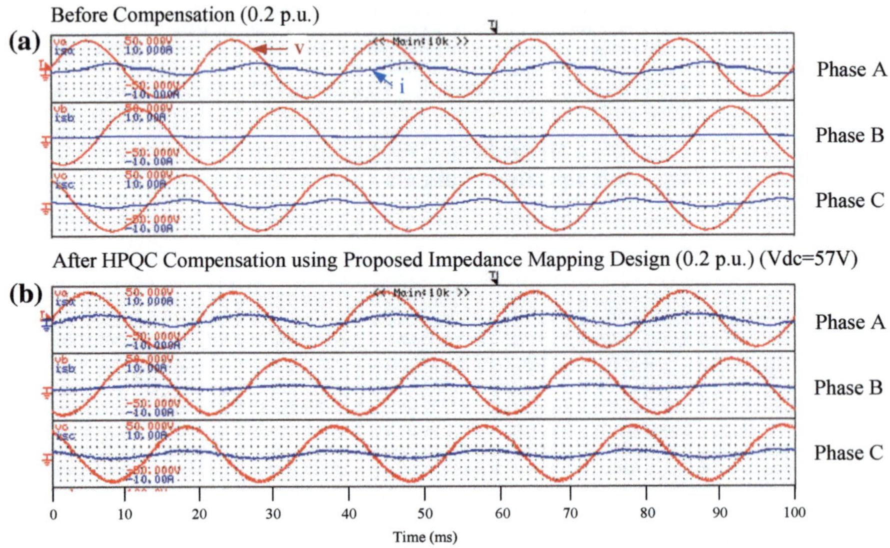

Fig. 4.22 Experimental waveforms obtained from hardware prototype of co-phase traction power supply with Railway HPQC under impedance-mapping design technique (load capacity = 0.2 p.u.)

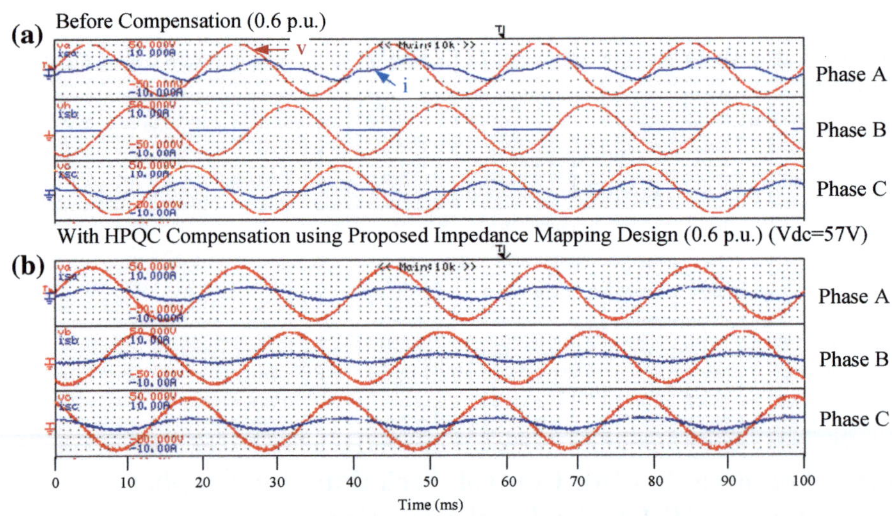

Fig. 4.23 Experimental waveforms obtained from hardware prototype of co-phase traction power supply with Railway HPQC under impedance-mapping design technique (load capacity = 0.6 p.u.)

load variation from rated value. This may cause additional power loss. For the second method, the coupled impedance is selected according to load variation range to reduce the operation voltage rating. However, as traction locomotive loadings are unpredictable and the transportation demand may increase due to country and area

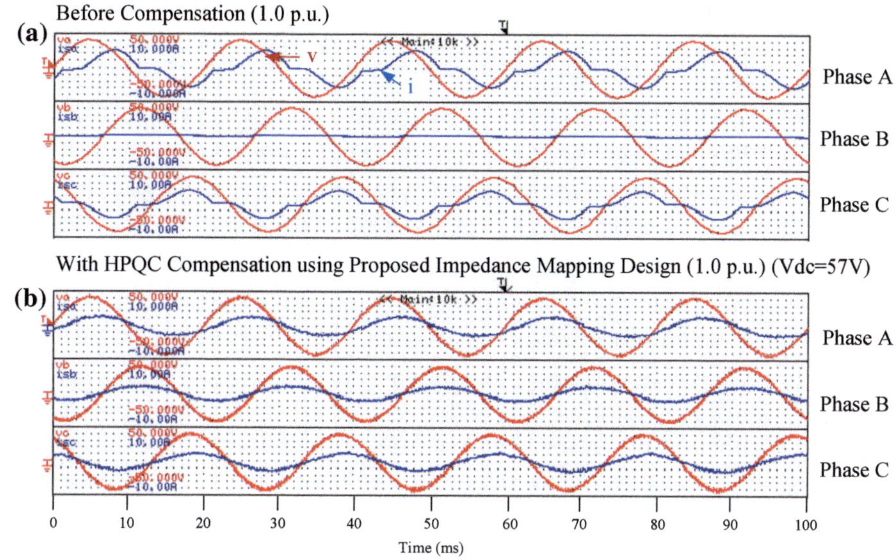

Fig. 4.24 Experimental waveforms obtained from hardware prototype of co-phase traction power supply with Railway HPQC under impedance-mapping design technique (load capacity = 1.0 p.u.)

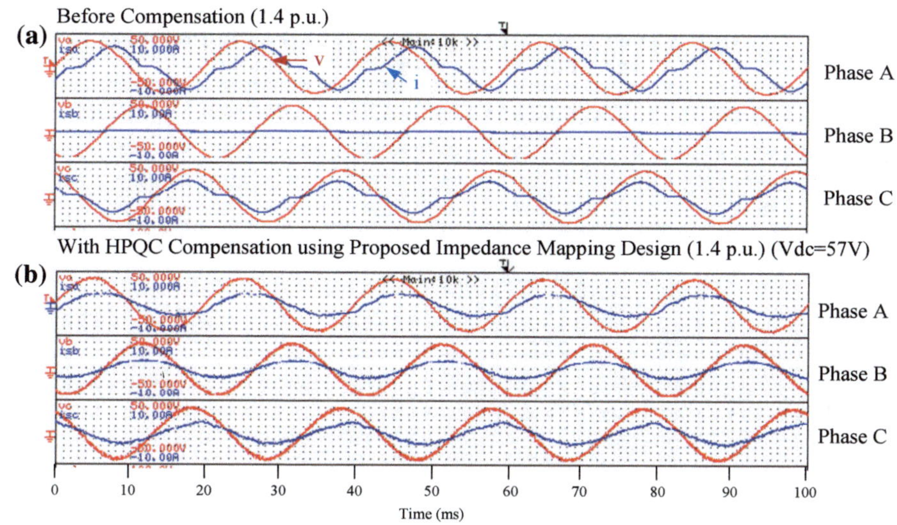

Fig. 4.25 Experimental waveforms obtained from hardware prototype of co-phase traction power supply with Railway HPQC under impedance-mapping design technique (load capacity = 1.4 p.u.)

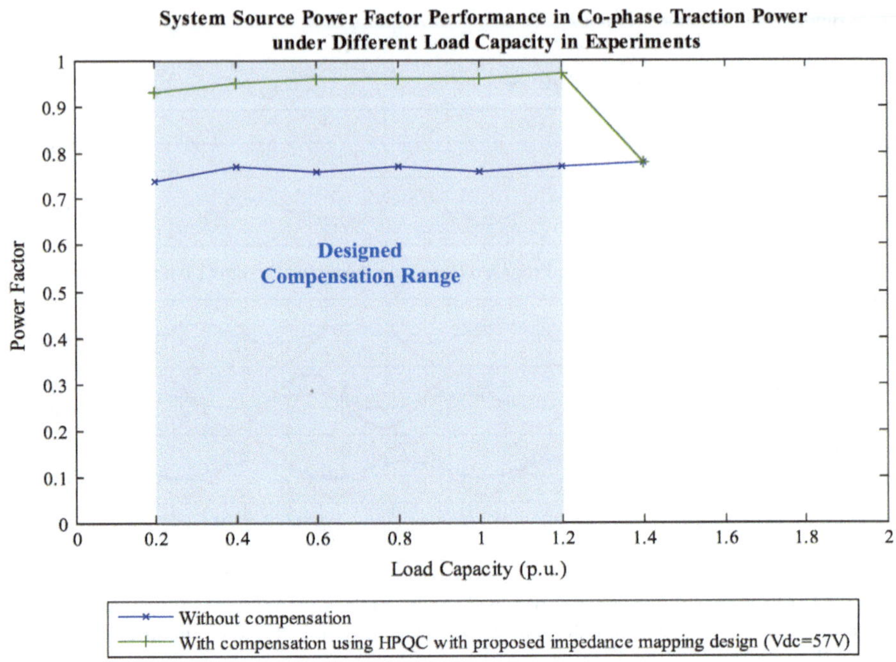

Fig. 4.26 Experimental data of system source power factor performance in co-phase traction power with Railway HPQC under impedance-mapping design under different load capacity

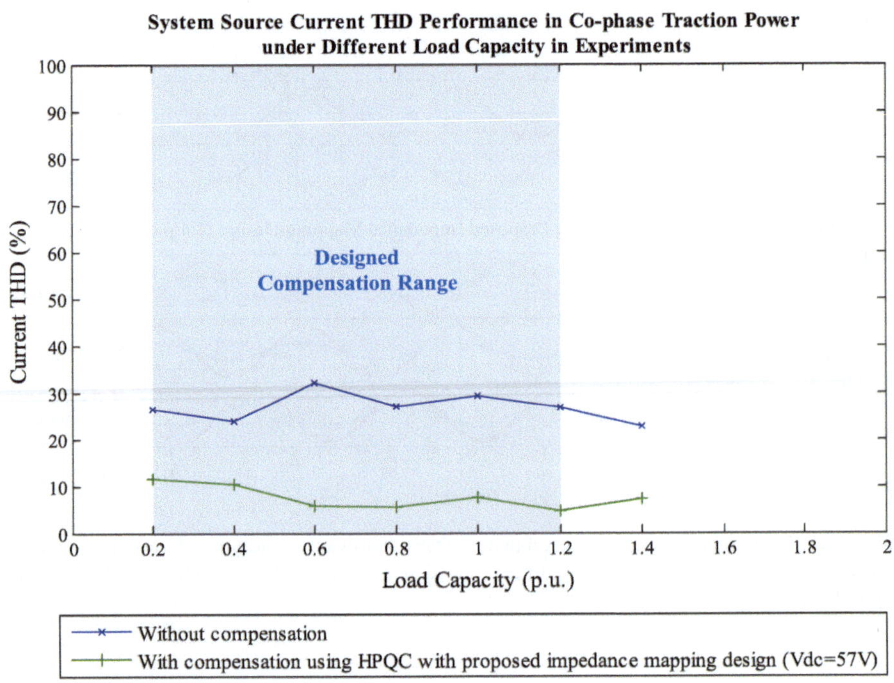

Fig. 4.27 Experimental data of system source current THD (%) in co-phase traction power with Railway HPQC under impedance-mapping design under different load capacity

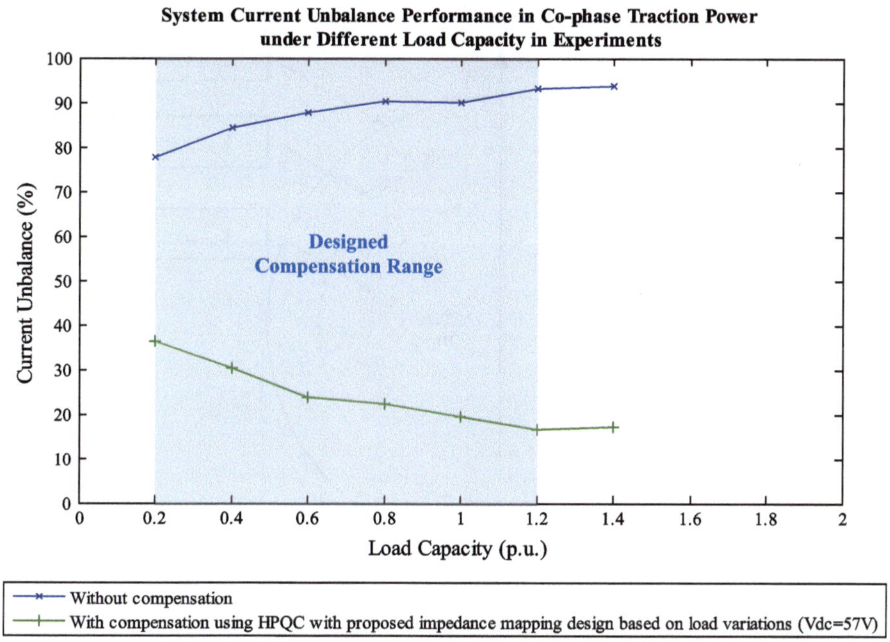

Fig. 4.28 Experimental data of system source current unbalance (%) in co-phase traction power with Railway HPQC under impedance-mapping design under different load capacity

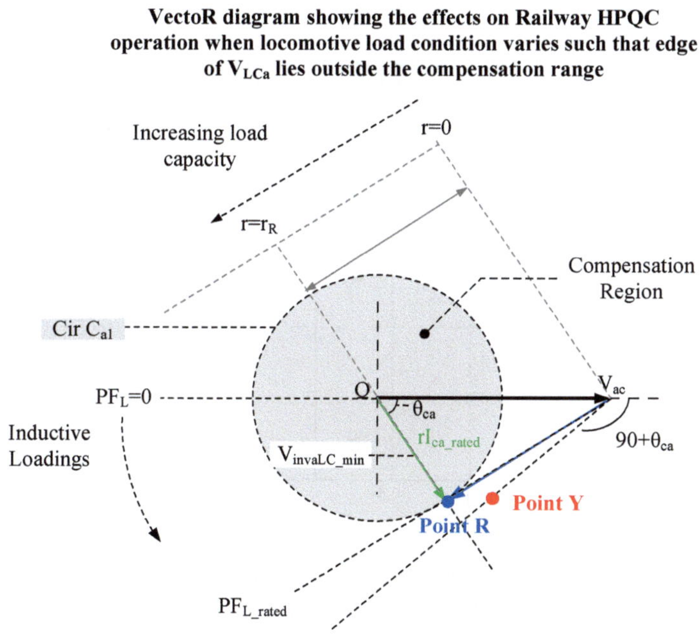

Fig. 4.29 Vector diagram showing the condition when the locomotive loading condition varies such that the edge of V_{LCa} in Railway HPQC lies outside the compensation range (point *Y*)

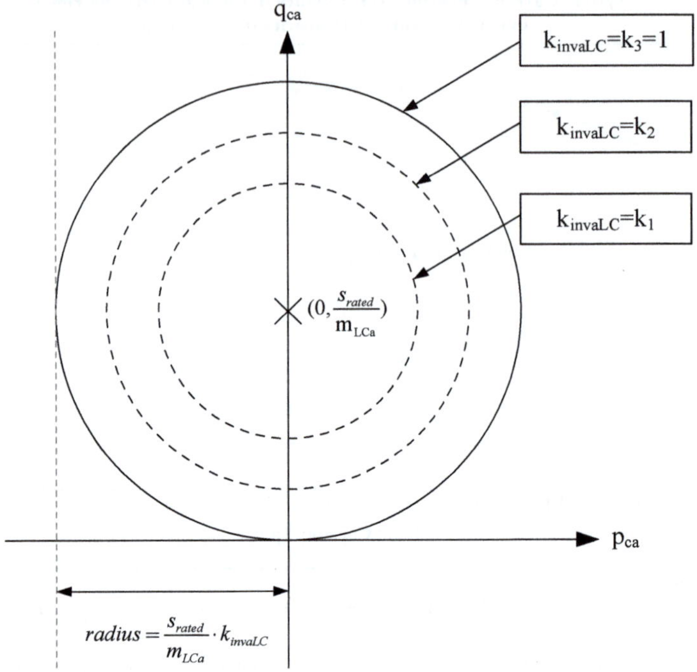

Fig. 4.30 A graph showing the direct relationship between Railway HPQC operation voltage rating and its active/reactive power output capability

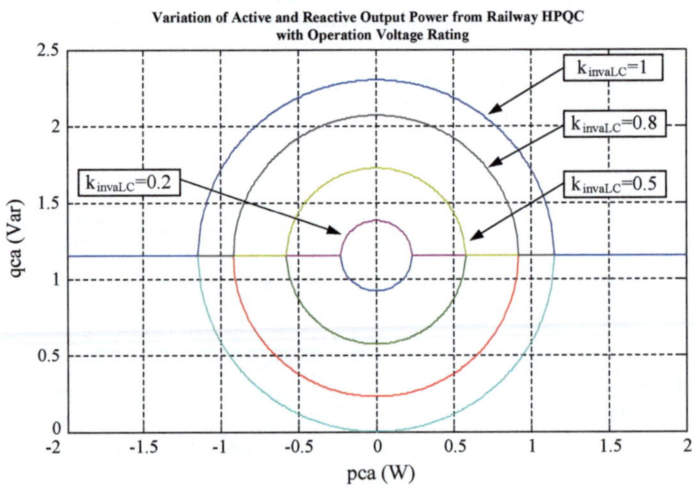

Fig. 4.31 A graph showing the variation of active and reactive output power from Railway HPQC with operation voltage rating as circles with the same origin

development, the operation voltage may need to be further enhanced no matter which Railway HPQC coupled impedance design is used. This may cause frequent manual operation during vigorous load variations. Based on this consideration, the adaptive dc link control in Railway HPQC is developed to automatically change the Railway HPQC operation voltage rating based on the load condition. This method is explored below.

4.5.1 Insufficient Operation Voltage of Railway HPQC When Load Varies

First of all, it should be noticed that unsatisfactory performance of Railway HPQC when load varies is due to insufficient operation voltage, which means that the Railway HPQC output capability is less than the requirement of compensation capability. This has already been discussed previously in Fig. 4.3 and will be further explored here.

Shown in Fig. 4.29 is a vector diagram showing a possible condition in Railway HPQC of co-phase traction power that the locomotive loading condition is varied such that the edge of vector V_{LCa} is located outside the compensation range. In the following, the location of the edge of V_{LCa} is regarded as the operation point. For example, when the Railway HPQC operation voltage is V_{invaLC_min}, the compensation range would be bounded by the circle with center at origin and radius of V_{invaLC_min}, shown as shaded region in the figure. Supposing that the Railway HPQC is designed based on the requirement of minimum operation voltage at rated load condition as in (3.10), satisfactory compensation performance can be provided only at rated load condition, indicated by point R in the figure. When the load condition is varied such that the edge of V_{LCa} is located at point Y, satisfactory compensation performance cannot be provided since point Y is no longer inside the shaded region. This can be understood in another way that the Railway HPQC power output capability (shaded compensation region) is less than the requirement of co-phase traction Railway HPQC power quality compensation power (the operation point Y). Therefore, in order to develop the adaptive dc link voltage control to adjust the operation voltage, the relationship between Railway HPQC output capability and power output requirement should be explored first.

4.5.2 Investigations of Relationship Between Railway HPQC Output Capability and Required Output Power

Railway HPQC Active and Reactive Power Output Capability

In order to show a clear picture between Railway HPQC operation voltage and power output capability (including active and reactive), more investigations are done based on the analysis above.

It is already mentioned that the Railway HPQC operation voltage is restricted by the expression in (4.6). For the parameter of coupled impedance X_{LCa}, it is referred to as the coupled impedance ratio which is expressed in terms of PCC voltage and rated compensation current, as in (2.42). The value of m_{LCa} can be designed based on different requirements, such as minimum operation voltage rating at rated load in (3.10), or based on impedance-mapping technique in (4.32).

In the analysis that follows, the rated compensation capacity s_{rated} is defined according to PCC voltage V_{ac} and rated compensation current I_{ca_rated}, as indicated in (4.51).

$$s_{rated} = V_{ac} I_{ca\_rated} \tag{4.51}$$

The active and reactive power output from Railway HPQC, p_{ca} and q_{ca}, is defined as presented in (4.52).

$$\begin{aligned} p_{ca} &= V_{ac} I_{cap} \\ q_{ca} &= V_{ac} I_{caq} \end{aligned} \tag{4.52}$$

By substituting (2.42), (4.51) and (4.52) into (4.6), the expressions in (4.53) can be obtained. The ratio of Railway HPQC operation voltage V_{invaLC} to PCC voltage V_{ac} is defined as operation voltage rating k_{invaLC}.

$$\begin{aligned} V_{invaLC} &\geq \sqrt{\left(V_{ac} - V_{ac} \cdot (V_{ac} I_{caq}) \cdot \frac{m_{LCa}}{V_{ac} I_{ca\_rated}}\right)^2 + \left(V_{ac} \cdot (V_{ac} I_{cap}) \cdot \frac{m_{LCa}}{V_{ac} I_{ca\_rated}}\right)^2} \\ \frac{V_{invaLC}}{V_{ac}} &\geq \sqrt{\left(1 - q_{ca} \cdot \frac{m_{LCa}}{s_{ca\_rated}}\right)^2 + \left(p_{ca} \cdot \frac{m_{LCa}}{s_{ca\_rated}}\right)^2} \\ k_{invaLC} &\geq \sqrt{\left(1 - q_{ca} \cdot \frac{m_{LCa}}{s_{ca\_rated}}\right)^2 + \left(p_{ca} \cdot \frac{m_{LCa}}{s_{ca\_rated}}\right)^2} \end{aligned} \tag{4.53}$$

By further manipulation of (4.53), the expression can be rearranged as shown in (4.54). The expression shows that the relationship between the Railway HPQC output active and reactive power capability, p_{ca} and q_{ca}, forms different circles under different operation voltage rating k_{invaLC}.

$$\left(\frac{p_{ca}}{\frac{s_{rated}}{m_{LCa}} \cdot k_{invaLC}}\right)^2 + \left(\frac{\frac{s_{rated}}{m_{LCa}} - q_{ca}}{\frac{s_{rated}}{m_{LCa}} \cdot k_{invaLC}}\right)^2 \leq 1 \tag{4.54}$$

It is mentioned above that the output active and reactive power, p_{ca} and q_{ca}, from the Railway HPQC is bounded by a circle. By investigating the expression in (4.54), the described circle is with origin at (0, s_{rated}/m_{LCa}), and radius of $s_{rated}/m_{LCa} \cdot k_{invaLC}$.

The relationship is shown graphically in Fig. 4.30. For instance, assuming traction load power factor value of 0.85, and Railway HPQC design for minimum operation voltage rating ($m_{\text{LCa}} = 0.87$), the variation of output active and reactive power with s_{rated} at 1.0 is shown in Fig. 4.31.

Many important points can be inferred from the figures, including, but not limited to the following:

- With a higher value of Railway HPQC operation voltage rating k_{invaLC}, more active and reactive power can be output;
- The middle point of the range in reactive power output is always located at $s_{\text{rated}}/m_{\text{LCa}}$;
- With value of Railway HPQC operation rating k_{invaLC} lower than 1.0, the Railway HPQC can only output inductive reactive power, not capacitive one;
- Given a fixed value of k_{invaLC}, the Railway HPQC can be used to inject as much active power as it can absorb; this is useful especially in bidirectional power flow;

Requirement of Railway HPQC Compensation Power Capability in Co-phase Traction Power

Next, the required active and reactive power output from the Railway HPQC for co-phase traction power quality is being discussed and analyzed. Since adaptive dc link control is mainly useful during load variations, this is done so to understand the required output characteristics when loading condition varies in order to determine the appropriate adaptive dc link control.

Referring to the compensation power requirement of co-phase traction power Railway HPQC in (2.17), the expressions in (4.55) can be obtained.

$$\begin{aligned} p_{\text{ca}} &= \frac{1}{2}(\text{PF}_L)(I_L)(V_{\text{ac}}) = \frac{1}{2}(\text{PF}_L)(s_L) \\ q_{\text{ca}} &= \left[\frac{1}{2\sqrt{3}}(\text{PF}_L) + \sqrt{1-(\text{PF}_L)^2}\right](I_L)(V_{\text{ac}}) = \left[\frac{1}{2\sqrt{3}}(\text{PF}_L) + \sqrt{1-(\text{PF}_L)^2}\right](s_L) \end{aligned} \tag{4.55}$$

By eliminating the value of s_L in (4.55), the relationship between the required active and reactive power output from Railway HPQC shown in (4.56) can be obtained.

$$\frac{q_{\text{ca}}}{p_{\text{ca}}} = \left\{ \frac{2 \cdot \left[\frac{1}{2\sqrt{3}}(\text{PF}_L) + \sqrt{1-(\text{PF}_L)^2}\right]}{\text{PF}_L} \right\} \tag{4.56}$$

It can be observed from the expression that the ratio of q_{ca} to p_{ca} is a function of load power factor PF_L. A Matlab plot is being constructed according to (4.56) and is shown in Fig. 4.32. The relationship between them is not linear. With load power

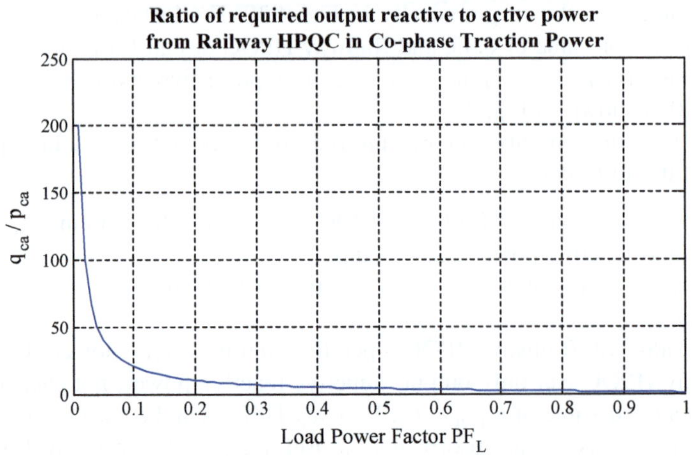

Fig. 4.32 A Matlab plot showing the variation of ratio between Railway HPQC output reactive and active power for co-phase traction power quality compensation with load power factor

factor = 0, the amount of required output reactive power q_{ca} is higher. As load power factor increases, the ratio decreases. For instance, for load power factor = 0.85, the ratio of q_{ca}/p_{ca} is 1.82.

Besides variation of load power factor, the load capacity may also vary during load variations. It has been analyzed previously that the compensation capacity is directly proportional to the load capacity. For example, when load capacity is changed from $s_{L\text{rated}}$ to r times of $s_{L\text{rated}}$, the required compensation capacity is also changed from s_{rated} to r times of s_{rated}. With the compensation capacity being a triangular relationship with output active and reactive power, the expression in (4.57) can be derived. It shows that the output compensation active and reactive power when load capacity varies would form a circle, with center at origin and radius of rs_{rated}.

$$(p_{ca})^2 + (q_{ca})^2 = (r \cdot s_{\text{rated}})^2 \tag{4.57}$$

The combination of the expressions in (4.56) and (4.57) under selected load power factor and capacity variation is plotted in Fig. 4.33. For load capacity variations, the condition is chosen from WuQing substation load capacity variation from 0.2 to 1.2 p.u. (with 0.6 p.u. as middle point and 1.5 p.u. for extraordinary case). It shows that the combination of Railway HPQC p_{ca} and q_{ca} in co-phase traction power quality compensation will rotate clockwise with increasing load power factor while the circle of possible p_{ca} and q_{ca} combination will expand with increasing load capacity. With fixed load power factor, the p_{ca} q_{ca} point would locate along a straight line; on the other hand, with fixed load capacity, the p_{ca} q_{ca} point would locate along the perimeter of the circle. Therefore, unlike active power filters for reactive power and harmonic compensation, given fixed load capacity and power factor, there is only one possible point of Railway HPQC active and reactive output power. For instance, with load power factor value of 0.85 and capacity value of 1.2 p.u., the only possible Railway HPQC active and reactive power output is located at point X (0.58, 1.06) in Fig. 4.33.

Further analysis on the relationship between Railway HPQC output capability and required output power

From the analysis above, it can be seen that although the Railway HPQC can output different combination of active and reactive power, the region of required one is different, the area of interest is located at the overlapping area of these two regions.

The graphics in Fig. 4.31 (Railway HPQC Output Capability) and Fig. 4.33 (Required Output Power) are combined and is shown in Fig. 4.34. It is supposed that the Railway HPQC is designed based on minimum operation voltage requirement under rated load power factor value of 0.85 ($m_{LCa} = 0.87$). For example, traction load power factor usually ranges from 0.8 to 0.9, the required output active/reactive power combination is located within the shaded region in the figure. When the load capacity is larger than the rated value, the required active/reactive compensation power gets larger and further away from the origin; on the other hand, when the load capacity is larger than rated value, the required active/reactive compensation power gets smaller and closer to the origin.

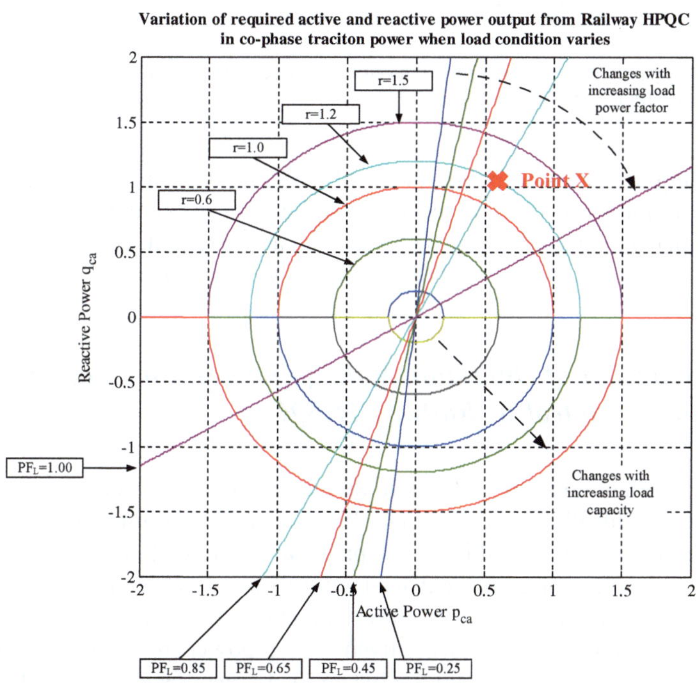

Fig. 4.33 Variation of required output active and reactive power from Railway HPQC for co-phase traction power quality compensation when load varies

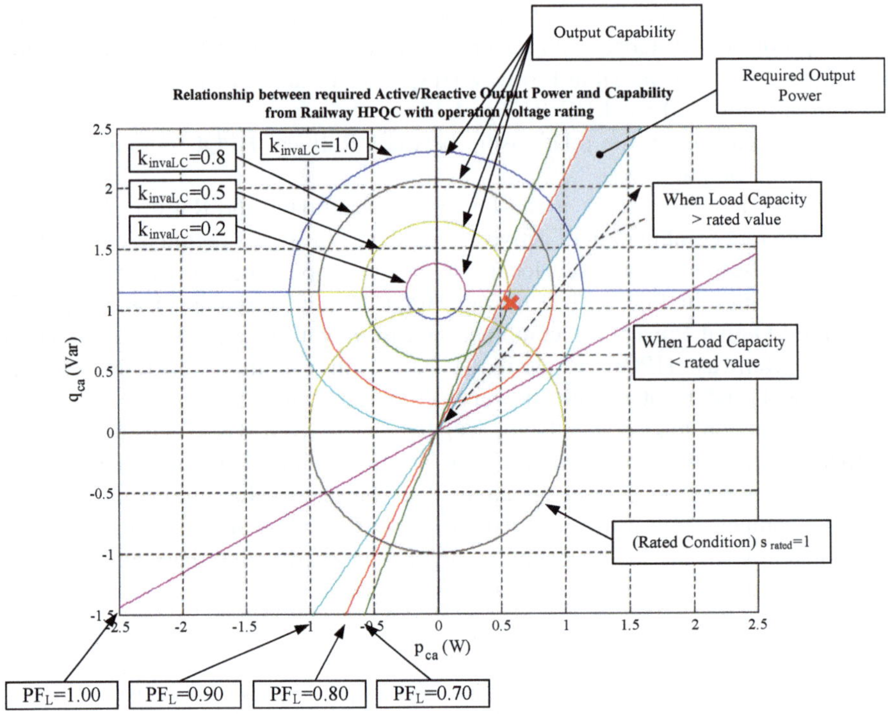

Fig. 4.34 A graphic plot showing the relationship between required active/reactive output power from Railway HPQC and output capability

For example, with load power factor of 0.85 and capacity of 1.2 times of rated value, the required *pq* point is located at point *X* in the figure. Satisfactory compensation performance can be provided as long as the region bounded by the output capability overlaps with that of the required power output.

4.5.3 Selection of Operation Voltage Region for Adaptive dc Link Control in Railway HPQC

Another important concern in adaptive dc link control is the selection of operation voltage region. Ideally, it is, of course, better to select the operation voltage from zero to infinity. However, this is not practical because it would consume more resources (for comparing and register memory) in controller like DSP or FPGA. Furthermore, it would be useless to have operation voltage region with too high or too low values since the required active/reactive power may not lie within it. For instance, there is hardly any condition when load power factor is zero. Therefore, it is preferable to determine the operation voltage region according to the range of loading conditions.

For a clearer explanation, a design example is given below. Based on the practical traction load data, the following design requirement can be deduced.

Rated Traction Load Capacity: 31.5 MVA;
Rated Traction Load Voltage: 27.5 kV;
Rated Railway HPQC Compensation Capacity: 27.83 MVA;
Traction Load Capacity Range: 6.3–37.8 MVA [0.2–1.2 p.u.]; and
Traction Load Power Factor Range: 0.7–1.0 (margin included).

For the adaptive dc link control design, the operation voltage range should be selected such that the Railway HPQC output capability can cover the required compensation range.

Supposing also that the Railway HPQC is designed based on minimum operation voltage rating at rated load power factor of 0.85 ($m_{LCa} = 0.87$), a Matlab plot constructed based on (4.53), (4.56), and (4.57) as well as the design requirement above is shown in Fig. 4.35. Detailed parameter values can be found in Table 4.6. For the value of k_{invaLC}, it is our determination target and will be covered below.

Based on the analysis above, the desired output compensation range is bounded by the two straight lines formed by load power factor range (PF_{L_H} and PF_{L_L}) as

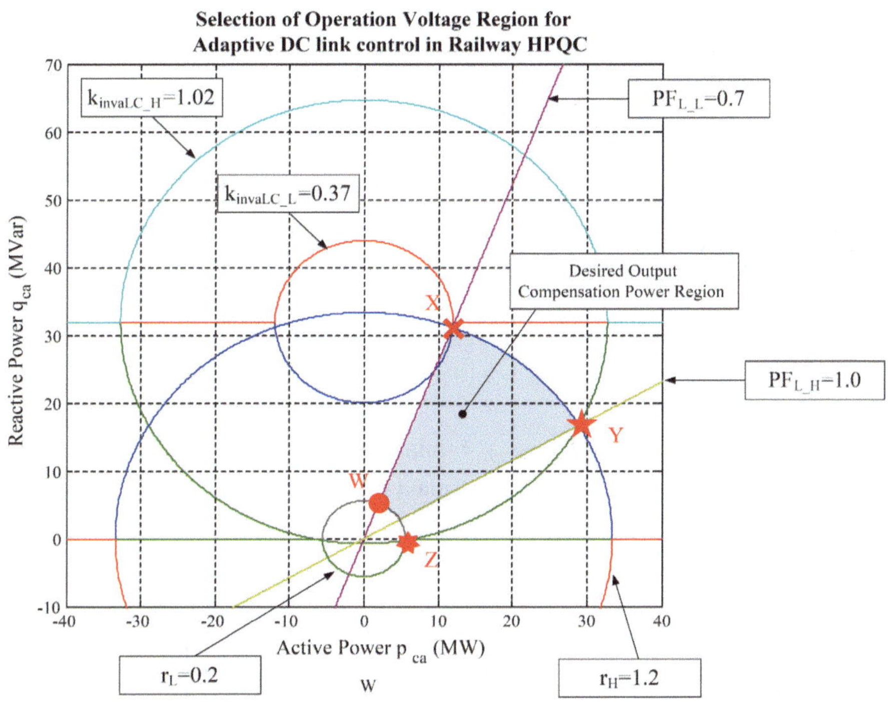

Fig. 4.35 A graphic plot showing the selection of operation voltage range in Railway HPQC adaptive dc link control so as to cover the desired output compensation power region

Table 4.6 Detailed parameters for plotting the graphs in Fig. 4.35

Parameters	Low value (min)	High value (max)
m_{LCa}	0.87	
s_{rated}	27.83 (MVA)	
r	$r_L = 0.2$	$r_H = 1.2$
PF_L	$\text{PF}_{\text{L\_L}} = 0.7$	$\text{PF}_{\text{L\_H}} = 1.0$

well as the circles formed by the load capacity range (r_H and r_L), shown as shaded region in Fig. 4.35. It is obvious that satisfactory compensation performance can be provided as long as the circle of the output capability can touch the boundary points (shown as points *W*, *X*, *Y*, and *Z* in the figure). The higher and lower values of k_{invaLCa} can, therefore, be determined by determining the solutions of (4.53), (4.56), and (4.57). The expressions derived for determination are shown in (4.58).

$$\begin{aligned}
W &= \sqrt{\left(1-(m_{\text{pqL}})\left(\frac{r_L\cdot m_{\text{LCa}}}{\sqrt{1+(m_{\text{pqL}})^2}}\right)\right)^2+\frac{(r_L\cdot m_{\text{LCa}})^2}{\sqrt{1+(m_{\text{pqL}})^2}}}\\
X &= \sqrt{\left(1-(m_{\text{pqL}})\left(\frac{r_H\cdot m_{\text{LCa}}}{\sqrt{1+(m_{\text{pqL}})^2}}\right)\right)^2+\frac{(r_H\cdot m_{\text{LCa}})^2}{\sqrt{1+(m_{\text{pqL}})^2}}}\\
Y &= \sqrt{\left(1-(m_{\text{pqH}})\left(\frac{r_H\cdot m_{\text{LCa}}}{\sqrt{1+(m_{\text{pqH}})^2}}\right)\right)^2+\frac{(r_H\cdot m_{\text{LCa}})^2}{\sqrt{1+(m_{\text{pqH}})^2}}}\\
Z &= \sqrt{\left(1-(m_{\text{pqH}})\left(\frac{r_H\cdot m_{\text{LCa}}}{\sqrt{1+(m_{\text{pqH}})^2}}\right)\right)^2+\frac{(r_H\cdot m_{\text{LCa}})^2}{\sqrt{1+(m_{\text{pqH}})^2}}}
\end{aligned} \tag{4.58}$$

$$\begin{cases} k_{\text{invaLC\_L}} = \min(W, X, Y, Z) \\ k_{\text{invaLC\_H}} = \max(W, X, Y, Z) \end{cases}$$

where

$$m_{\text{pqL}} = \left\{\frac{2\cdot\left[\frac{1}{2\sqrt{3}}(\text{PF}_{L\_L})+\sqrt{1-(\text{PF}_{L\_L})^2}\right]}{\text{PF}_{L\_L}}\right\}$$

and

$$m_{\mathrm{pqH}} = \left\{ \frac{2 \cdot \left[\frac{1}{2\sqrt{3}} (\mathrm{PF}_{L\_H}) + \sqrt{1 - (\mathrm{PF}_{L\_H})^2} \right]}{\mathrm{PF}_{L\_H}} \right\}$$

By substituting the design requirements in the example, the values calculated are $k_{\mathrm{invaLC\_L}} = 0.37$ and $k_{\mathrm{invaLC\_H}} = 1.02$. The circles of Railway HPQC output capability under these two conditions are also shown in Fig. 4.35. It shows that when the operation voltage region is selected according to these values, the Railway HPQC can provide the desired output active and reactive power within the load variation range.

The higher ($k_{\mathrm{invaLC\_H}}$) and lower ($k_{\mathrm{invaLC\_L}}$) operation voltage rating as defined in (4.59).

$$V_{\mathrm{dc\_interval}} = \sqrt{2} \cdot \left(\frac{k_{\mathrm{invaLC\_H}} - k_{\mathrm{invaLC\_L}}}{n} \right) \tag{4.59}$$

4.5.4 Modification of Adaptive dc Link Voltage Control in Railway HPQC Control Algorithm

So far, the development of adaptive dc link voltage control is focused on the required operation voltage rating but has little relationship with the computation and control block of Railway HPQC.

As introduced previously, instantaneous *pq* theory for single phase is used for computation in the control of Railway HPQC. Clarke transform is used and it is important to investigate the relationship between the parameters in the instantaneous *pq* theory and the parameters in the determination of Railway HPQC operation voltage rating in (4.53). Furthermore, it is also important that the adaptive dc link voltage control can be combined with existing Railway HPQC computation block in (2.17). These two issues are mainly discussed below.

Relationship between instantaneous *pq* theory and traditional active/reactive power

First of all, the relationship between the power in instantaneous *pq* theory and traditional active/reactive power is being investigated. Detailed computation and control block diagram of Railway HPQC in co-phase traction power supply system is shown in Fig. 2.7.

The locomotive load active/reactive power is determined by the single-phase instantaneous *pq* theory in (2.26). The expressions for instantaneous active and reactive power computation is shown again in (4.60). In order to distinguish, p_{inst} and q_{inst} are used to define instantaneous active and reactive power.

$$\begin{aligned} p_L &= p_{\text{inst}} = v_\alpha i_\alpha + v_\beta i_\beta \\ q_L &= q_{\text{inst}} = v_\beta i_\alpha - v_\alpha i_\beta \end{aligned} \tag{4.60}$$

The definition of v_α, v_β, i_α, and i_β may be referred to (2.24) and (2.25). v_β and i_β are 90° delay of v_α and i_α, respectively. It is further assumed that v_α, v_β, i_α, and i_β are defined as in (4.61) and (4.62).

$$\begin{cases} v_\alpha = v_{\text{ac}} = \sqrt{2}V_{\text{ac}} \sin(\omega t) \\ v_\beta = v_{\text{ac}} e^{-j\pi/2} = \sqrt{2}V_{\text{ac}} \cos(\omega t) \end{cases} \tag{4.61}$$

$$\begin{cases} i_\alpha = i_L = \sqrt{2}I_L \sin(\omega t + \phi_L) \\ i_\beta = i_L e^{-j\pi/2} = \sqrt{2}I_L \cos(\omega t + \phi_L) \end{cases} \tag{4.62}$$

By substituting (4.61) and (4.62) into (4.60) and further manipulations, the expressions in (4.63) can be obtained.

$$\begin{cases} p_{\text{inst}} = 2V_{\text{ac}} I_L \cos\phi_L = 2p_{\text{trad}} \\ q_{\text{inst}} = 2V_{\text{ac}} I_L \sin\phi_L = 2q_{\text{trad}} \end{cases} \tag{4.63}$$

Therefore, the traditional active and reactive power may be determined from instantaneous active and reactive power using (4.64).

$$\begin{cases} p_{\text{trad}} = \dfrac{p_{\text{inst}}}{2} \\ q_{\text{trad}} = \dfrac{q_{\text{inst}}}{2} \end{cases} \tag{4.64}$$

Therefore, the expression for determination of required Railway HPQC operation voltage rating developed for adaptive dc link control in (4.53) can be revised as in (4.65), where $p_{\text{ca\_inst}}$ and $q_{\text{ca\_inst}}$ are the instantaneous active and reactive compensation power computed as in the control block diagram of Fig. 2.7.

$$\begin{aligned} k_{\text{invaLC}} &\geq \sqrt{\left(1 - q_{\text{ca\_trad}} \cdot \frac{m_{\text{LCa}}}{S_{\text{ca\_rated}}}\right)^2 + \left(p_{\text{ca\_trad}} \cdot \frac{m_{\text{LCa}}}{S_{\text{ca\_rated}}}\right)^2} \\ &\geq \sqrt{\left(1 - \frac{q_{\text{ca\_inst}}}{2} \cdot \frac{m_{\text{LCa}}}{S_{\text{ca\_rated}}}\right)^2 + \left(\frac{p_{\text{ca\_inst}}}{2} \cdot \frac{m_{\text{LCa}}}{S_{\text{ca\_rated}}}\right)^2} \end{aligned} \tag{4.65}$$

Modified Control Block Diagram

As shown in Fig. 4.36 is the modified control block diagram of Railway HPQC control algorithm including adaptive dc link voltage control. The modified portions are shown as a red shaded block. There are mainly two important issues,

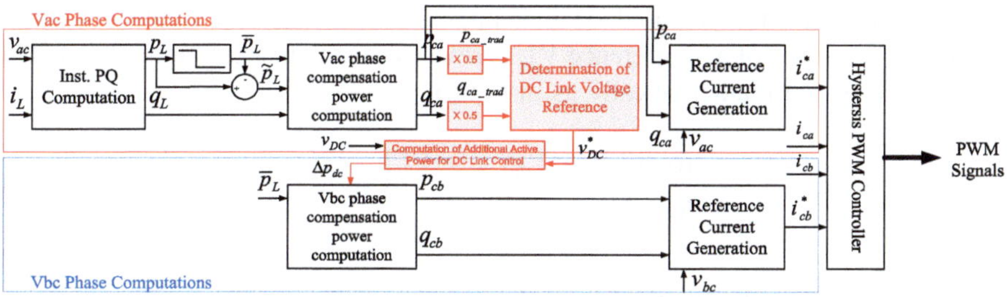

Fig. 4.36 Modified control block diagram of Railway HPQC control algorithm including adaptive dc link voltage control

namely, (1) determination of dc voltage reference, and (2) computation of required additional active power for dc link voltage control. Each of them is discussed below.

A. Determination of dc link voltage reference V_{dc}^{*}

As discussed, the required dc link voltage can be determined from required operation voltage rating. The expression for determining required operation voltage rating of Railway HPQC based on loading conditions is shown in (4.65).

$$V_{dc\_req} = \sqrt{2} k_{invaLC} \geq \sqrt{2\left(1 - \frac{q_{ca\_inst}}{2} \cdot \frac{m_{LCa}}{s_{ca\_rated}}\right)^2 + 2\left(\frac{p_{ca\_inst}}{2} \cdot \frac{m_{LCa}}{s_{ca\_rated}}\right)^2} \tag{4.66}$$

In adaptive dc link voltage control, it is important that the dc link voltage reference is determined so that the dc link voltage is controlled to the reference level. Although the dc link voltage reference may be adjusted at every sample cycle according to load conditions, however, this is not preferred since sudden load condition variation may cause rapid dc link voltage reference changes and causes frequent power adjustment, which may degrade Railway HPQC compensation performance. Therefore, adaptive dc link voltage level control method is developed. As shown in (4.59), the dc link voltage interval is selected according to the highest and lowest possible dc link voltage. The dc link voltage interval may be determined by the expression in (4.67).

$$\begin{aligned} V_{dc}^{*}[i] &= \sqrt{2}(k_{invaLC\_L} + (i-1) \cdot V_{dc\_interval}) \\ &= \sqrt{2} \cdot \left(k_{invaLC\_L} + (i-1) \cdot \frac{k_{invaLC\_H} - k_{invaLC\_L}}{n}\right), \\ &\text{for } 1 \leq i \leq n \end{aligned} \tag{4.67}$$

Supposing that the required dc link operation voltage is computed based on (4.66), the dc link voltage reference V_{dc}^* is selected as the nearest higher level. This may be expressed mathematically as in (4.68).

$$\begin{aligned} &\text{for } V_{dc}^*[k] < V_{dc\_req} < V_{dc}^*[k+1], \\ &\text{then } V_{dc}^* = V_{dc}^*[k+1] \end{aligned} \tag{4.68}$$

B. Computation of required additional active power for dc link voltage control

Besides the determination of dc link voltage reference in (4.68), the dc link voltage is to be controlled at the reference level. In order to increase or decrease the Railway HPQC dc link voltage according to the reference level, additional active power control is required to absorb/release active power from or into the system. For the two converters in Railway HPQC, this additional power is suggested to be absorbed/released by the V_{bc} phase converter, which is connected to the unloaded phase. This method has following advantages:

- Save dc Link Voltage even when V_{ac} phase converter fails
  Active and reactive power are absorbed from the system through V_{bc} phase converter during compensation (notice the negative sign of p_{cb} and q_{cb} in Table 2.4); by doing dc link voltage control using V_{bc} phase converter, the dc link voltage can be controlled even when V_{ac} phase converter operation fails (on the other hand, if V_{ac} phase converter is used to do absorb power for dc link voltage control, when V_{ac} phase converter operation fails, large amount of active power will be absorbed from V_{bc} phase converter and is stored in the dc link voltage, which causes excessively high dc link voltage and is dangerous;
- Less influence on compensation performance on locomotive load
  Since V_{ac} phase converter is connected across locomotive load and is used to compensate locomotive load reactive power and harmonics, if dc link voltage control is accomplished by V_{ac} phase converter, the additional power may influence the compensation performance on locomotive load

Therefore, when adaptive dc link voltage control is included, the compensation control algorithm in (2.17) is revised as in (4.69). Additional active power portion ΔP_{dc} are added to the V_{bc} phase active compensation power. The additional portion ΔP_{dc} is determined by the difference with actual dc link voltage V_{dc} and dc link voltage reference V_{dc}^*, using proportional P control.

$$\begin{bmatrix} p_{ca} \\ q_{ca} \\ p_{cb} \\ q_{cb} \end{bmatrix} = \begin{bmatrix} K_1\bar{p}_L + \tilde{p}_L \\ K_2\bar{p}_L + q_L \\ K_3\bar{p}_L + \Delta P_{dc} \\ K_4\bar{p}_L \end{bmatrix} = \begin{bmatrix} K_1\bar{p}_L + \tilde{p}_L \\ K_2\bar{p}_L + q_L \\ K_3\bar{p}_L + P\left(V_{dc} - V_{dc}^*\right) \\ K_4\bar{p}_L \end{bmatrix} \tag{4.69}$$

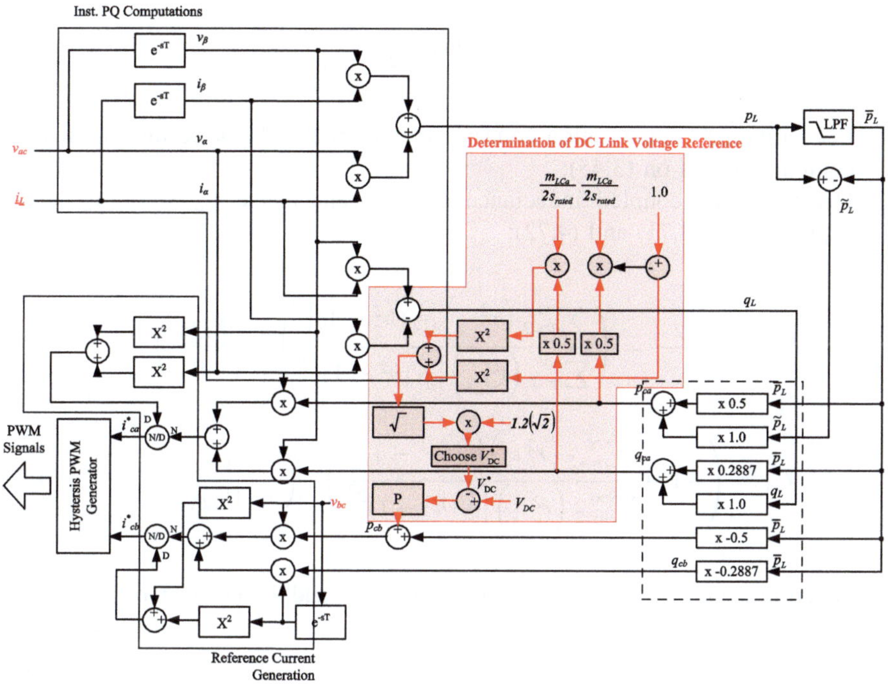

Fig. 4.37 Modified detailed control block diagram including adaptive dc link voltage control

Based on the discussion above, the Railway HPQC detailed control block diagram in Fig. 2.9 is revised as in Fig. 4.37. The modified parts about determination of dc link voltage reference are shown as shaded red region in the figure based on (4.66) and (4.69).

4.5.5 Comprehensive Design Procedures for Co-phase Railway HPQC with Adaptive dc Link Control Technique

The comprehensive design procedure for Railway HPQC based on impedance-mapping technique is shown below:

1. Obtain and compute the value of locomotive PCC voltage V_{ac}, rated compensation current I_{ca_rated}, and V_{ac} phase compensation angle θ_{ca};
2. Determine the value of V_{ac} phase coupled impedance ratio m_{LCa} (any);
3. Select the value of V_{ac} phase coupled impedance X_{LCa} according to (4.70);

$$|X_{\mathrm{LCa}}| = m_{\mathrm{LCa}} \cdot \frac{V_{\mathrm{ac}}}{I_{\mathrm{ca\_rated}}} \tag{4.70}$$

4. For harmonic compensation concern, the coupled inductance ratio k_L may be computed based on (3.45);
5. Determine the coupled inductance L_a and capacitance C_a according to the expression in (4.71) and (4.72);

$$L_a = \frac{k_L |X_{\mathrm{LCa}}|}{\omega} = \left(\frac{\sum_{h=2}^{\infty} \left(r_{h\_h}^2 \right) \left[\frac{1}{h^2} (h^2 - 1) \right]}{\sum_{h=2}^{\infty} \left(r_{h\_h}^2 \right) \left[\frac{1}{h^2} (h^2 - 1)^2 \right]} \right) \cdot \left(\frac{|X_{\mathrm{LCa}}|}{\omega} \right) \tag{4.71}$$

$$C_a = \left(-1 - \frac{\sum_{h=2}^{\infty} \left(r_{h\_h}^2 \right) \left[\frac{1}{h^2} (h^2 - 1) \right]}{\sum_{h=2}^{\infty} \left(r_{h\_h}^2 \right) \left[\frac{1}{h^2} (h^2 - 1)^2 \right]} \right) \cdot \left(\frac{1}{\omega \cdot |X_{\mathrm{LCa}}|} \right) \tag{4.72}$$

6. Determine the V_{bc} phase coupled impedance and inductance by (4.73) and (4.74);

$$|X_{\mathrm{LCb}}| \geq \frac{V_{\mathrm{bc}} \sin \theta_{\mathrm{cb}} - \sqrt{V_{\mathrm{invaLC\_min}}^2 - V_{\mathrm{bc}}^2 \cos^2 \theta_{\mathrm{cb}}}}{I_{\mathrm{cb}}} \tag{4.73}$$

$$L_b = \frac{|X_{\mathrm{LCb}}|}{\omega} \tag{4.74}$$

7. Determine the load capacity range in per unit value, and define the lower one as r_A and upper one as r_B;
8. Compute the active and reactive variation ratio h_c and h_s by (4.26);
9. Compute and determine the operation voltage rating range of Railway HPQC, $k_{\mathrm{invaLC\_L}}$ to $k_{\mathrm{invaLC\_H}}$, according to (4.58);

$$h_c = \frac{\cos \theta_{\mathrm{ca}}}{\cos \theta_{\mathrm{ca\_rated}}} \qquad h_s = \frac{\sin \theta_{\mathrm{ca}}}{\sin \theta_{\mathrm{ca\_rated}}} \tag{4.75}$$

HPQC, $k_{\mathrm{invaLC\_L}}$ to $\mathrm{k}_{\mathrm{invaLC\_H}}$, according to (4.58);

$$W = \sqrt{\left(1 - (m_{\mathrm{pqL}})\left(\frac{r_L \cdot m_{\mathrm{LCa}}}{\sqrt{1 + (m_{\mathrm{pqL}})^2}}\right)\right)^2 + \frac{(r_L \cdot m_{\mathrm{LCa}})^2}{\sqrt{1 + (m_{\mathrm{pqL}})^2}}}$$

$$X = \sqrt{\left(1 - (m_{\mathrm{pqL}})\left(\frac{r_H \cdot m_{\mathrm{LCa}}}{\sqrt{1 + (m_{\mathrm{pqL}})^2}}\right)\right)^2 + \frac{(r_H \cdot m_{\mathrm{LCa}})^2}{\sqrt{1 + (m_{\mathrm{pqL}})^2}}}$$

$$Y = \sqrt{\left(1 - (m_{\mathrm{pqH}})\left(\frac{r_H \cdot m_{\mathrm{LCa}}}{\sqrt{1 + (m_{\mathrm{pqH}})^2}}\right)\right)^2 + \frac{(r_H \cdot m_{\mathrm{LCa}})^2}{\sqrt{1 + (m_{\mathrm{pqH}})^2}}}$$

$$Z = \sqrt{\left(1 - (m_{\mathrm{pqH}})\left(\frac{r_L \cdot m_{\mathrm{LCa}}}{\sqrt{1 + (m_{\mathrm{pqH}})^2}}\right)\right)^2 + \frac{(r_L \cdot m_{\mathrm{LCa}})^2}{\sqrt{1 + (m_{\mathrm{pqH}})^2}}}$$

$$\begin{cases} k_{\mathrm{invaLC\_L}} = \min(W, X, Y, Z) \\ k_{\mathrm{invaLC\_H}} = \max(W, X, Y, Z) \end{cases}$$

where

$$m_{\mathrm{pqL}} = \left\{ \frac{2 \cdot \left[\frac{1}{2\sqrt{3}}(\mathrm{PF}_{L\_L}) + \sqrt{1 - (\mathrm{PF}_{L\_L})^2}\right]}{\mathrm{PF}_{L\_L}} \right\}$$

and

$$m_{\mathrm{pqH}} = \left\{ \frac{2 \cdot \left[\frac{1}{2\sqrt{3}}(\mathrm{PF}_{L\_H}) + \sqrt{1 - (\mathrm{PF}_{L\_H})^2}\right]}{\mathrm{PF}_{L\_H}} \right\}$$

10. Compute the Railway HPQC dc Link Voltage interval (*n* intervals) according to (4.76);

$$V_{\mathrm{dc\_interval}} = \sqrt{2} \cdot \left(\frac{k_{\mathrm{invaLC\_H}} - k_{\mathrm{invaLC\_L}}}{n}\right) \tag{4.76}$$

11. Modify the control block diagram of Railway HPQC according to Fig. 4.37 and (4.68)

4.5.6 Simulation Study

Similar to previous analysis, simulations are performed using PSCAD. The circuit schematic of co-phase traction power supply system for simulation verifications are shown in Fig. 3.15. In order to verify the adaptive dc link voltage control performance during load variations, to different loading conditions are used, namely, Load 1 (Rated: 1.0 p.u., 31.5 MVA) and Load 2 (Half Rated: 0.5 p.u., 15.75 MVA). Load 1 is set to be connected to the system at 0.2 s, while load 1 is changed to load 2 at 0.5 s.

The Railway HPQC coupled impedance is determined according to (3.51). The design procedures of adaptive dc link voltage control are shown in Sect. 4.5.5. The Railway HPQC parameter used are shown in Table 4.7.

The determination parameters for the dc link operation voltage range of Railway HPQC for load variations in simulations are shown in Table 4.8.

The simulated system source current and voltage waveforms are shown in Figs. 4.38, 4.39, 4.40, and 4.41. Shown in Fig. 4.38 are the simulated load current, source current, and dc link voltage waveforms during the simulation. As

Table 4.7 System parameters used in the simulation verification of Railway HPQC using adaptive dc link voltage control for load variations

No.	Parameters	Symbol	Value
1.	V_{ac} phase coupled impedance ratio	m_{LCa}	0.8761
2.	V_{ac} phase coupled inductance ratio	k_L	0.0912
3.	V_{ac} phase coupled inductance	L_a	6.5 mH
4.	V_{ac} phase coupled capacitance	C_a	131 μF
5.	V_{bc} phase coupled inductance	L_b	6 mH
6.	dc link capacitance	C_{dc}	10,000 μF
7.	Railway HPQC operation voltage rating range	k_{invaLC}	0.37–1.02
8.	dc link operation voltage range	V_{dc}	14.5–39.8 kV

Table 4.8 Determination for dc link operation voltage range in Railway HPQC of co-phase traction for load variations

No.	Parameters	Symbol	Value
1.	Minimum load capacity	r_A	0.2
2.	Maximum load capacity	r_B	1.2
3.	Lower load power factor	PF_{L_L}	0.7
4.	Higher load power factor	PF_{L_H}	1.0
5.	Higher dc link operation voltage rating	k_{invaLC_H}	1.02
6.	Lower dc link operation voltage rating	k_{invaLC_L}	0.37
7.	Higher dc link operation voltage	V_{dc_H}	14.5 kV
8.	Lower dc link operation voltage	V_{dc_L}	39.8 kV
9.	dc link operation voltage interval	$V_{dc\_interval}$	8.4 kV

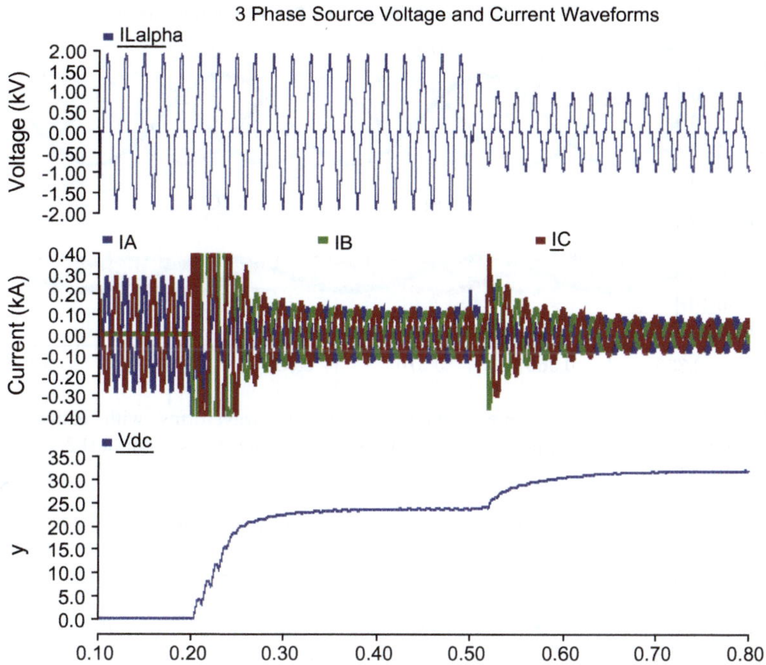

Fig. 4.38 Simulated load current, source current, and dc link voltage waveforms for adaptive dc link voltage control verification

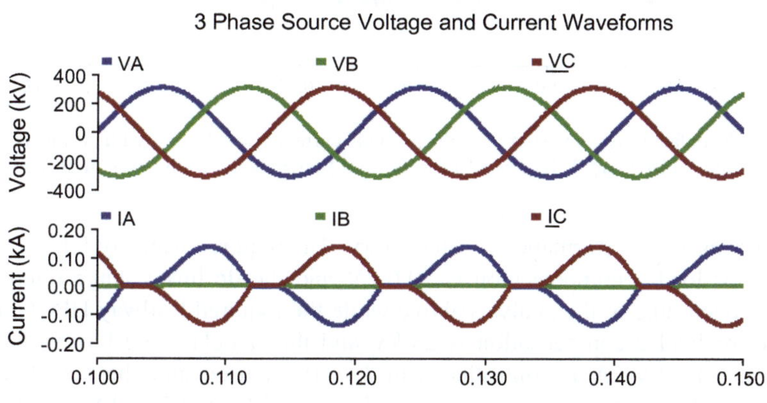

Fig. 4.39 Simulated system source voltage and current waveforms without any compensation

introduced, the load 1 (rated 1.0 p.u.) is connected at 0.2 s while load 2 (half rated 0.5 p.u.) is connected at 0.5 s. It can be observed that the amplitude of the load current is reduced at 0.5 s. It can be also observed that even when load condition is changed, the Railway HPQC can adaptively change its operation voltage such that

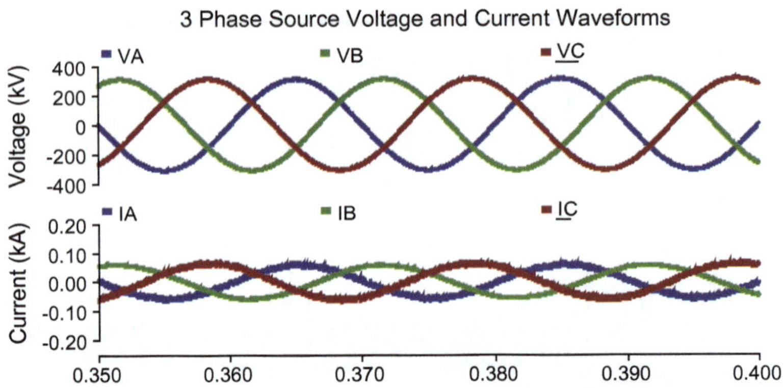

Fig. 4.40 Simulated system source voltage and current waveforms with Railway HPQC Compensation under Load 1 (rated load, 1.0 p.u.) (switched into the system at 0.2 s)

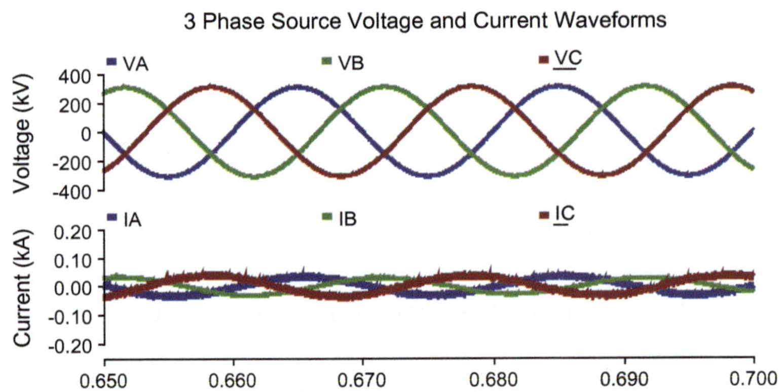

Fig. 4.41 Simulated system source voltage and current waveforms with Railway HPQC Compensation under Load 2 (half-rated load, 0.5 p.u.) (switched into the system at 0.5 s)

the compensation performance is satisfactory. The required Railway HPQC dc link voltage for load 1 compensation is 20.6 kV and the dc link voltage reference is 23.3 kV according to the analysis above while the required Railway HPQC dc link voltage for load 2 compensation is 28 kV and the calculated dc link voltage reference is 31.5 kV. A P controller with $P = 10$ is used and the actual dc link operation voltage during compensation for load 1 is 23.5 and 31.9 kV, respectively. The error between the actual Railway HPQC dc link voltage and reference one is 0.9 and 1.3%, respectively.

For reference, the simulated voltage and current waveforms without any compensation are given in Fig. 4.38. It can be seen that without compensation, the system suffers from high harmonic distortions, reactive power and system unbalance. When the Railway HPQC compensator is put into operation at 0.2 s, the dc link voltage increases and is controlled to a certain level such that satisfactory compensation performance can be provided. The steady-state performance for load 1 under compensation is shown in Fig. 4.40. The power quality problems are being compensated. The source current is harmonic free and balanced; the source reactive power is also eliminated and reduced. When the load is changed to load 2 (half rated 0.5 p.u.) at 0.5 s, the dc link voltage is adjusted again such that satisfactory compensation performance can be provided. The steady-state source voltage and current waveforms under load 2 condition with Railway HPQC compensation are shown in Fig. 4.41. It can be observed that the source current is also harmonic free, system unbalance and reactive power is also reduced.

In order to show clearly the system performance, the detailed power quality data, namely, system source power factor, source current total harmonic distortions, and system unbalance are shown in Table 4.9. It can be observed that.

4.5.7 Experimental Results

Experimental results are obtained to verify the performance of co-phase traction power supply with adaptive dc link control. The Railway HPQC parameters are designed according to the procedures in Sect. 4.5.5, and are shown in Table 4.10. It

Table 4.9 Simulated system performance data for the adaptive dc link voltage control verifications

	V_{dc}^{*}	V_{dc}	PF	THD (%)	I_{un} (%)
Without compensation (before 0.2 s)	–	–	0.60	27.78	99
Load 1 (1.0 p.u.) (0.2–0.5 s)	23.3 kV	23.5 kV	0.99	3.15	6.80
Load 2 (0.5 p.u.) (after 0.5 s)	31.5 kV	32 kV	0.99	2.69	13.32

Table 4.10 Experimental parameter design of the Railway HPQC under adaptive dc link control

No.	Parameters	Symbol	Value
1.	V_{ac} phase coupled impedance ratio	m_{LCa}	0.876
2.	V_{ac} phase coupled inductance ratio	k_L	0.10
3.	V_{ac} phase coupled inductance	L_a	4.9 mH
4.	V_{ac} phase coupled capacitance	C_a	170 μF
5.	V_{bc} phase coupled inductance	L_b	4 mH
6.	dc link capacitance	C_{dc}	10,000 μF
7.	Railway HPQC operation voltage rating	k_{invaLC}	0.37–1.02
8.	dc link operation voltage	V_{dc}	26–72 V
9.	Number of adaptive dc link intervals	n	3
10.	dc link voltage levels	V_{DCn}	26 V 41.7 V 57 V 72 V

is designed based on the assumption that the traction load capacity changes from 0.2 to 1.2 p.u., while traction lower power factor changes from 0.7 to 1.0.

In the experiment, the load power factor is changed from PF = 0.87 to PF = 1.0 around rated load capacity. The system waveforms obtained are shown in Figs. 4.42, 4.43, and 4.44. The change of dc link voltage from 41 to 72 V when load power factor changes are captured in Fig. 4.42. The steady-state system waveforms of the conditions when PF = 0.87 and PF = 1.0 are presented in Figs. 4.43 and 4.44. The detailed power quality data and screens captured from Power Quality Analyzer are shown in Figs. 4.45, 4.46, 4.47, and 4.48. It can be seen that the dc link voltage is automatically changed to a higher level when the load power factor changes to a higher level. Therefore, by using adaptive dc link voltage control, the Railway HPQC can operate at a relatively low operation voltage level when load changes.

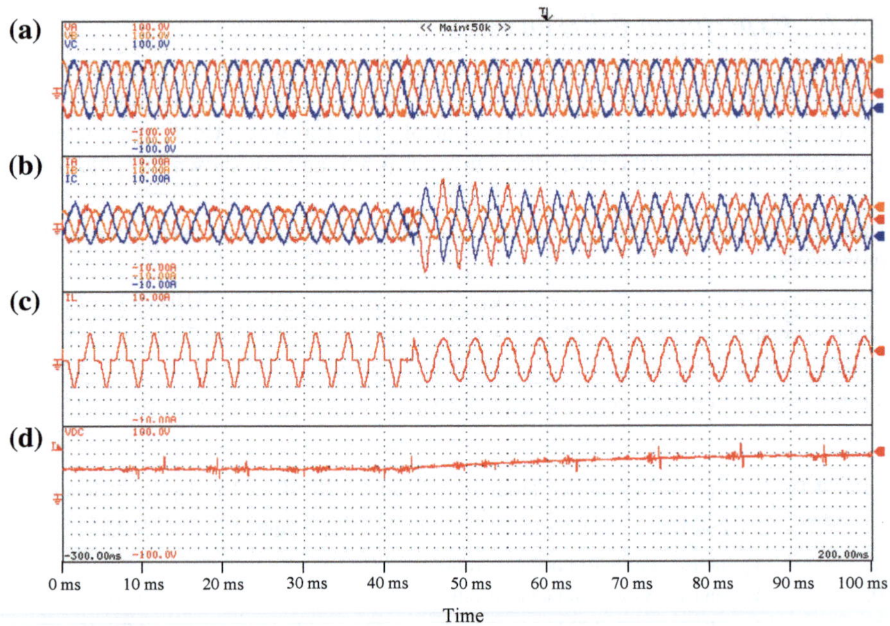

Fig. 4.42 Experimental system waveforms of co-phase traction power with Railway HPQC using adaptive dc link voltage control when traction load power factor increases from 0.87 to 1.0. **a** Three-phase source grid voltage; **b** three-phase source grid current; **c** load current; **d** Railway HPQC dc link voltage

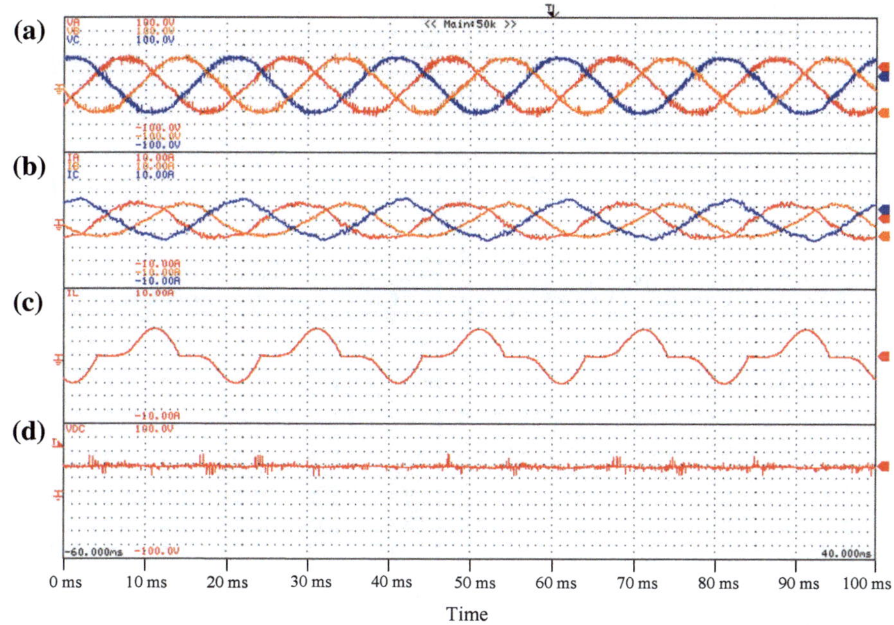

Fig. 4.43 Experimental system waveforms of co-phase traction power with Railway HPQC using adaptive dc link voltage control (PF = 0.87, V_{dc} = 41 V). **a** Three-phase source grid voltage; **b** three-phase source grid current; **c** load current; **d** Railway HPQC dc link voltage

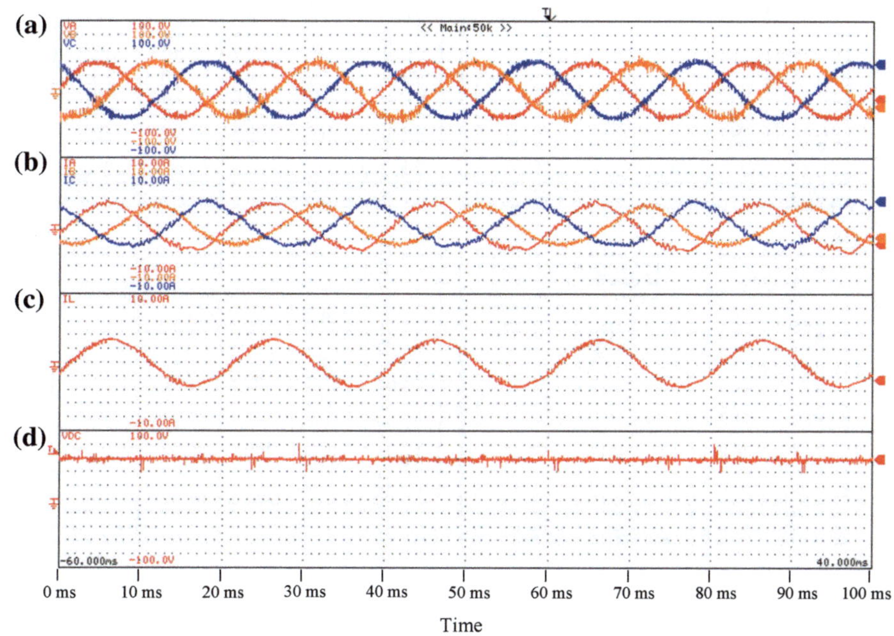

Fig. 4.44 Experimental system waveforms of co-phase traction power with Railway HPQC using adaptive dc link voltage control (PF = 1.0, V_{dc} = 72 V). **a** Three-phase source grid voltage; **b** three-phase source grid current; **c** load current; **d** Railway HPQC dc link voltage

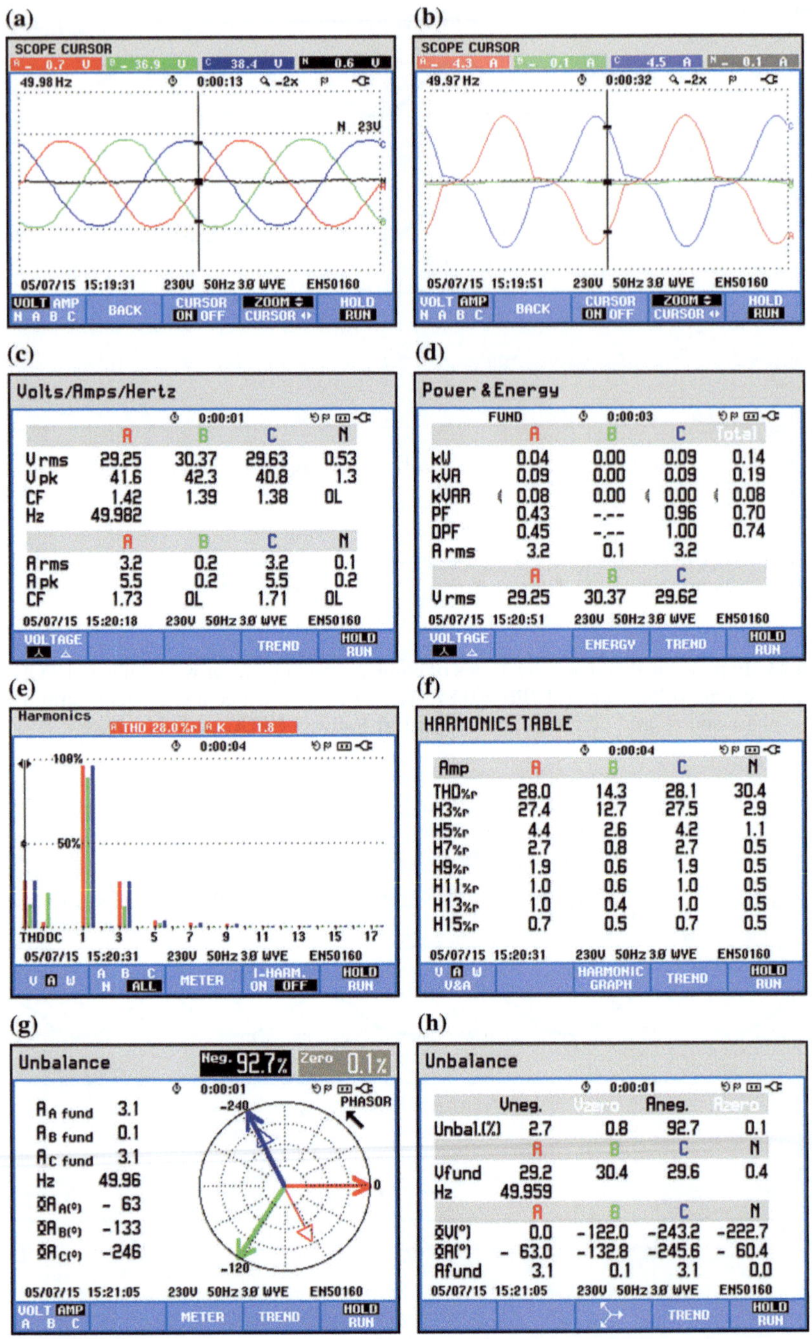
(a)
SCOPE CURSOR
A 0.7 V B 36.9 V C 38.4 V N 0.6 V
49.98 Hz 0:00:13 -2x
N 23V
05/07/15 15:19:31 230V 50Hz 3Ø WYE EN50160
VOLT AMP N A B C BACK CURSOR ON OFF ZOOM CURSOR HOLD RUN
(b)
SCOPE CURSOR
A 4.3 A B 0.1 A C 4.5 A N 0.1 A
49.97 Hz 0:00:32 -2x
05/07/15 15:19:51 230V 50Hz 3Ø WYE EN50160
VOLT AMP N A B C BACK CURSOR ON OFF ZOOM CURSOR HOLD RUN
(c)
Volts/Amps/Hertz
0:00:01
A B C N
Vrms 29.25 30.37 29.63 0.53
Vpk 41.6 42.3 40.8 1.3
CF 1.42 1.39 1.38 OL
Hz 49.982
A B C N
Arms 3.2 0.2 3.2 0.1
Apk 5.5 0.2 5.5 0.2
CF 1.73 OL 1.71 OL
05/07/15 15:20:18 230V 50Hz 3Ø WYE EN50160
VOLTAGE TREND HOLD RUN
(d)
Power & Energy
FUND 0:00:03
A B C Total
kW 0.04 0.00 0.09 0.14
kVA 0.09 0.00 0.09 0.19
kVAR 0.08 0.00 0.00 0.08
PF 0.43 -.-- 0.96 0.70
DPF 0.45 -.-- 1.00 0.74
Arms 3.2 0.1 3.2
A B C
Vrms 29.25 30.37 29.62
05/07/15 15:20:51 230V 50Hz 3Ø WYE EN50160
VOLTAGE ENERGY TREND HOLD RUN
(e)
Harmonics
A THD 28.0%r A K 1.8
0:00:04
100%
50%
THD DC 1 3 5 7 9 11 13 15 17
05/07/15 15:20:31 230V 50Hz 3Ø WYE EN50160
V A W A B C N ALL METER I-HARM. ON OFF HOLD RUN
(f)
HARMONICS TABLE
0:00:04
Amp A B C N
THD%r 28.0 14.3 28.1 30.4
H3%r 27.4 12.7 27.5 2.9
H5%r 4.4 2.6 4.2 1.1
H7%r 2.7 0.8 2.7 0.5
H9%r 1.9 0.6 1.9 0.5
H11%r 1.0 0.6 1.0 0.5
H13%r 1.0 0.4 1.0 0.5
H15%r 0.7 0.5 0.7 0.5
05/07/15 15:20:31 230V 50Hz 3Ø WYE EN50160
V A W V&A HARMONIC GRAPH TREND HOLD RUN
(g)
Unbalance
Neg. 92.7% Zero 0.1%
0:00:01
PHASOR
AA fund 3.1
AB fund 0.1
AC fund 3.1
Hz 49.96
ΦAA(°) - 63
ΦAB(°) -133
ΦAC(°) -246
-240
0
-120
05/07/15 15:21:05 230V 50Hz 3Ø WYE EN50160
VOLT AMP A B C METER TREND HOLD RUN
(h)
Unbalance
0:00:01
Vneg. Vzero Aneg. Azero
Unbal.(%) 2.7 0.8 92.7 0.1
A B C N
Vfund 29.2 30.4 29.6 0.4
Hz 49.959
A B C N
ΦV(°) 0.0 -122.0 -243.2 -222.7
ΦA(°) - 63.0 -132.8 -245.6 - 60.4
Afund 3.1 0.1 3.1 0.0
05/07/15 15:21:05 230V 50Hz 3Ø WYE EN50160
TREND HOLD RUN

◂**Fig. 4.45** Experimental results of screens captured from hardware prototype of co-phase traction power without compensation (PF = 0.87): **a** source voltage waveform; **b** source current waveform; **c** three-phase source voltage and current statistics; **d** three-phase source power and power factor; **e** source current harmonics spectrum; **f** three-phase source current harmonics statistics; **g** source current negative sequence; **h** source voltage and current unbalance data

4.6 Comparisons Among Different Railway HPQC Design for Load Variations

So far, three different methods are introduced for co-phase traction Railway HPQC operation under load variations. The three different methods all have their own advantages and disadvantages. Comparisons between the three different methods and criteria for determining preferred method are discussed in this section.

4.6.1 Enhancing Railway HPQC Compensation Capability by Increasing Operation Voltage

As the name implies, this method increases the Railway HPQC compensation capability by enhancing operation voltage. This method benefits from that changes in Railway HPQC coupled impedance is not required. Usually, when the co-phase traction power and Railway HPQC is installed, changes in the topology or coupled impedance value may be done so when the system is disconnected and stopped functioning, which will affect the system operation. However, since the operation voltage is increased, the operation loss is also increased. Therefore, this method is preferred only when the increase in operation voltage is not much. The analysis here acts as preliminary analysis for the relationship between Railway HPQC operation voltage and compensation capability.

4.6.2 Impedance-Mapping Technique According to Load Variation Range

In this method, the Railway HPQC coupled impedance is determined according to load variation range. It may be beneficial for reduction in operation voltage and coupled capacitance. However, it is important to determine the condition first to see if the operation voltage and coupled impedance can be reduced. This may be done so by checking if the conditions in (4.37) and (4.41) are satisfied. This method is preferred only if the conditions are satisfied.

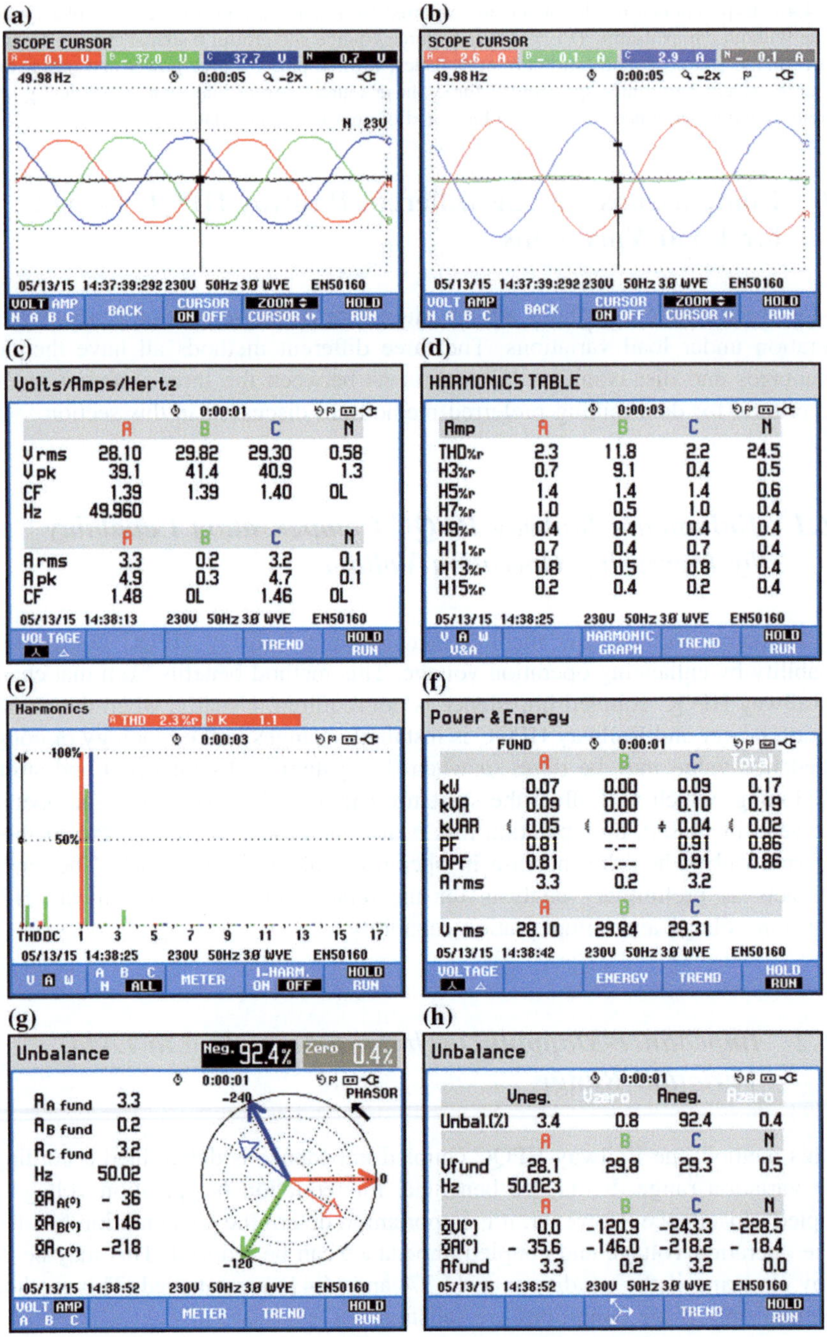
(a)
SCOPE CURSOR
49.98 Hz
0:00:05
N 230V
05/13/15 14:37:39:292 230V 50Hz 3Ø WYE EN50160
(b)
SCOPE CURSOR
49.98 Hz
0:00:05
05/13/15 14:37:39:292 230V 50Hz 3Ø WYE EN50160
(c)
Volts/Amps/Hertz
0:00:01
A B C N
Vrms 28.10 29.82 29.30 0.58
Vpk 39.1 41.4 40.9 1.3
CF 1.39 1.39 1.40 OL
Hz 49.960
A B C N
Arms 3.3 0.2 3.2 0.1
Apk 4.9 0.3 4.7 0.1
CF 1.48 OL 1.46 OL
05/13/15 14:38:13 230V 50Hz 3Ø WYE EN50160
(d)
HARMONICS TABLE
0:00:03
Amp A B C N
THD%r 2.3 11.8 2.2 24.5
H3%r 0.7 9.1 0.4 0.5
H5%r 1.4 1.4 1.4 0.6
H7%r 1.0 0.5 1.0 0.4
H9%r 0.4 0.4 0.4 0.4
H11%r 0.7 0.4 0.7 0.4
H13%r 0.8 0.5 0.8 0.4
H15%r 0.2 0.4 0.2 0.4
05/13/15 14:38:25 230V 50Hz 3Ø WYE EN50160
(e)
Harmonics
A THD 2.3%r
A K 1.1
0:00:03
100%
50%
THD DC 1 3 5 7 9 11 13 15 17
05/13/15 14:38:25 230V 50Hz 3Ø WYE EN50160
(f)
Power & Energy
FUND
0:00:01
A B C Total
kW 0.07 0.00 0.09 0.17
kVA 0.09 0.00 0.10 0.19
kVAR 0.05 0.00 0.04 0.02
PF 0.81 -.-- 0.91 0.86
DPF 0.81 -.-- 0.91 0.86
Arms 3.3 0.2 3.2
A B C
Vrms 28.10 29.84 29.31
05/13/15 14:38:42 230V 50Hz 3Ø WYE EN50160
(g)
Unbalance
Neg. 92.4%
Zero 0.4%
0:00:01
PHASOR
A A fund 3.3
A B fund 0.2
A C fund 3.2
Hz 50.02
ØA A(°) - 36
ØA B(°) -146
ØA C(°) -218
-240
0
-120
05/13/15 14:38:52 230V 50Hz 3Ø WYE EN50160
(h)
Unbalance
0:00:01
Vneg. Vzero Aneg. Azero
Unbal.(%) 3.4 0.8 92.4 0.4
A B C N
Vfund 28.1 29.8 29.3 0.5
Hz 50.023
A B C N
ØV(°) 0.0 -120.9 -243.3 -228.5
ØA(°) - 35.6 -146.0 -218.2 - 18.4
Afund 3.3 0.2 3.2 0.0
05/13/15 14:38:52 230V 50Hz 3Ø WYE EN50160

◀**Fig. 4.46** Experimental results of screens captured from hardware prototype of co-phase traction power without compensation (PF = 1.00): **a** source voltage waveform; **b** source current waveform; **c** three-phase source voltage and current statistics; **d** three-phase source power and power factor; **e** source current harmonics spectrum; **f** three-phase source current harmonics statistics; **g** source current negative sequence; **h** source voltage and current unbalance data

4.6.3 Adaptive dc Link Control Technique

Finally, adaptive dc link control technique is developed in Railway HPQC of co-phase traction power. The dc link voltage is changed automatically according to the load condition. This method is beneficial that none of the hardware topology and parameters need to be changed. Modifications are mainly done on the control (software) only. However, this control method is not suitable for vigorous load variations, which will lead to rapid changes in the dc link voltage and cause unstable system performances. Moreover, this adaptive dc link control technique may not be suitable for PWM control that is sensitive to dc link voltage such as voltage-controlled PWM.

4.7 Summary

In this chapter, three different methods are introduced to design co-phase traction power supply with Railway HPQC for locomotive loadings that vary within a certain range. As the locomotive load condition changes, the amount of required output compensation active and reactive power by Railway HPQC also changes. Under the design developed for minimum operation voltage at fixed rated in Chap. 3, the compensation capability is limited. Therefore, certain techniques are required and they are briefly discussed, analyzed, and compared in this chapter.

The first method is to increase the compensation capability of Railway HPQC in co-phase traction power by increasing the operation voltage. The relationship between Railway HPQC compensation capability and operation voltage is explored to develop the design. It is shown through simulation and experimental results that the system performance is satisfactory within the load variation range. This method is beneficial that no hardware parameter needs to be changed, but the operation voltage and coupled capacitance is not minimized. Therefore, it is suitable for conditions when changes in hardware are not preferred.

The second method is to apply the newly investigated impedance-mapping technique by our group. As the name implies, the method works by mapping the coupled impedance with the load/compensation power variation range. This helps to reduce the operation voltage and coupled impedance while providing similar system performance within load variation range. This method is advantageous for reduced operation voltage and coupled capacitance, as well as the cost. However, this method is suitable for the design stage of Railway HPQC when hardware

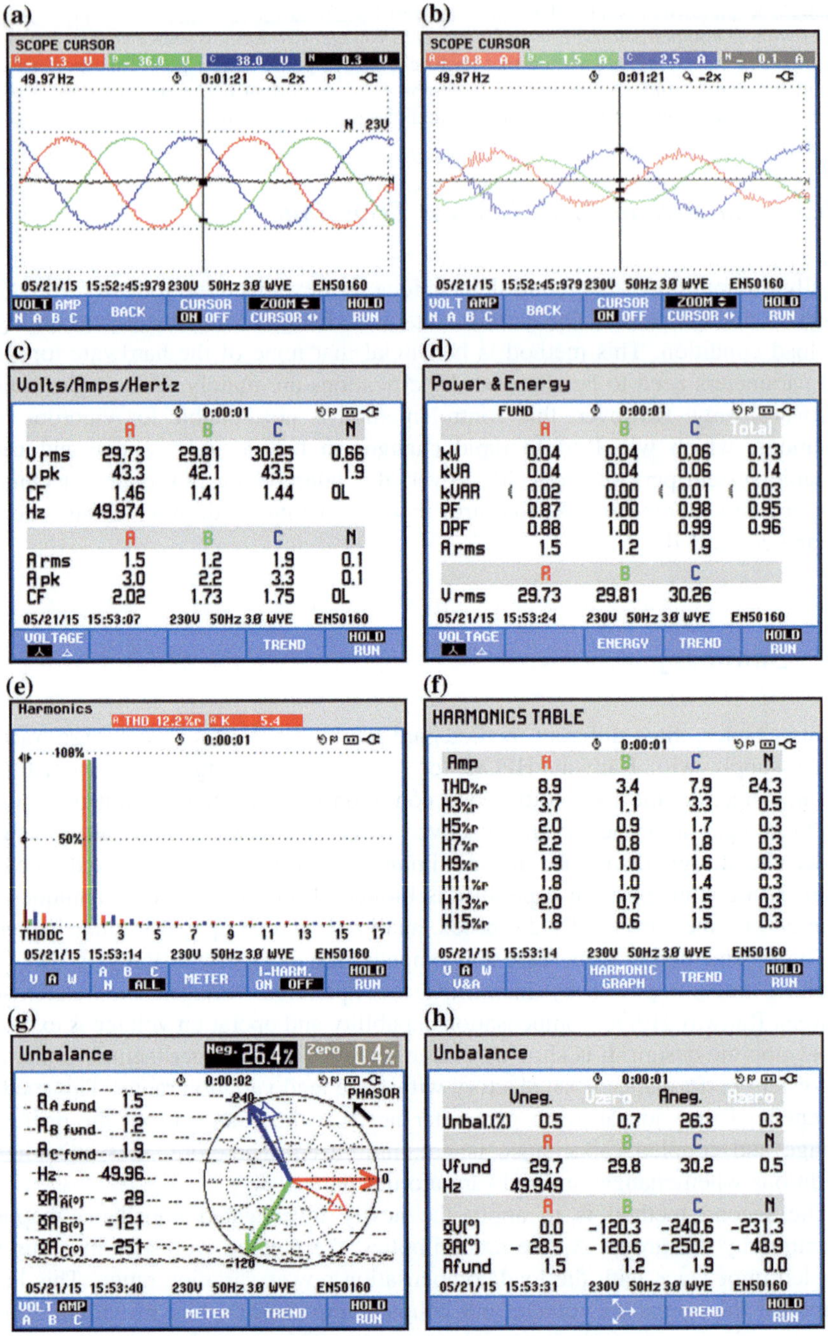
(a)
SCOPE CURSOR
49.97 Hz 0:01:21
05/21/15 15:52:45:979 230V 50Hz 3Ø WYE EN50160
(b)
SCOPE CURSOR
49.97 Hz 0:01:21
05/21/15 15:52:45:979 230V 50Hz 3Ø WYE EN50160
(c)
Volts/Amps/Hertz
0:00:01
A B C N
Vrms 29.73 29.81 30.25 0.66
Vpk 43.3 42.1 43.5 1.9
CF 1.46 1.41 1.44 OL
Hz 49.974
A B C N
Arms 1.5 1.2 1.9 0.1
Apk 3.0 2.2 3.3 0.1
CF 2.02 1.73 1.75 OL
05/21/15 15:53:07 230V 50Hz 3Ø WYE EN50160
(d)
Power & Energy
FUND 0:00:01
A B C Total
kW 0.04 0.04 0.06 0.13
kVA 0.04 0.04 0.06 0.14
kVAR 0.02 0.00 0.01 0.03
PF 0.87 1.00 0.98 0.95
DPF 0.88 1.00 0.99 0.96
Arms 1.5 1.2 1.9
A B C
Vrms 29.73 29.81 30.26
05/21/15 15:53:24 230V 50Hz 3Ø WYE EN50160
(e)
Harmonics
THD 12.2 %r K 5.4
0:00:01
100%
50%
THDDC 1 3 5 7 9 11 13 15 17
05/21/15 15:53:14 230V 50Hz 3Ø WYE EN50160
(f)
HARMONICS TABLE
0:00:01
Amp A B C N
THD%r 8.9 3.4 7.9 24.3
H3%r 3.7 1.1 3.3 0.5
H5%r 2.0 0.9 1.7 0.3
H7%r 2.2 0.8 1.8 0.3
H9%r 1.9 1.0 1.6 0.3
H11%r 1.8 1.0 1.4 0.3
H13%r 2.0 0.7 1.5 0.3
H15%r 1.8 0.6 1.5 0.3
05/21/15 15:53:14 230V 50Hz 3Ø WYE EN50160
(g)
Unbalance
Neg. 26.4% Zero 0.4%
0:00:02
PHASOR
A A fund 1.5
A B fund 1.2
A C fund 1.9
Hz 49.96
-240
0
-120
05/21/15 15:53:40 230V 50Hz 3Ø WYE EN50160
(h)
Unbalance
0:00:01
Vneg. Vzero Aneg. Azero
Unbal.(%) 0.5 0.7 26.3 0.3
A B C N
Vfund 29.7 29.8 30.2 0.5
Hz 49.949
A B C N
ΦV(°) 0.0 -120.3 -240.6 -231.3
ΦA(°) -28.5 -120.6 -250.2 -48.9
Afund 1.5 1.2 1.9 0.0
05/21/15 15:53:31 230V 50Hz 3Ø WYE EN50160

◂**Fig. 4.47** Experimental results of screens captured from hardware prototype of co-phase traction power with Railway HPQC compensation under new harmonic filter design (PF = 0.87, V_{dc} = 41 V): a)source voltage waveform; b) source current waveform; c) three-phase source voltage and current statistics; d) three-phase source power and power factor; e) source current harmonics spectrum; f) three-phase source current harmonics statistics; g) source current negative sequence; h) source voltage and current unbalance data

parameters can be changed. In the future, when impedance changing technology is achievable easily and widely used, this method will bring large advantages in reducing Railway HPQC operation voltage.

The third method is to apply the adaptive dc link voltage control technique to automatically adjust the operation voltage. By doing so, the Railway HPQC operation voltage can be reduced or minimized even when load condition changes. As active power transfer is involved during compensation, the design is different from traditional one which reactive power compensation is mainly concerned. The relationship between Railway HPQC operation voltage and compensation capability has already been analyzed in the first method. In order to avoid rapid changes in the operation voltage, the dc link voltage is divided into different levels and the criteria for determining whether changes operation voltage are required is also developed. The adaptive dc link voltage control is mainly accomplished by modifying the software programming in control. This method is beneficial that no hardware changes are required and can be considered as an advanced method of the first method (increasing operation voltage). However, this method is not suitable for conditions where load condition varies rapidly, as this will cause rapid changes in the dc link operation voltage requirement and will make the system unstable. Moreover, this method is also not suitable when the PWM technique is highly sensitive to changes in dc link voltage.

For reference, the design flow chart of the three different methods are shown in Figs. 4.49, 4.50, and 4.51. There is no so-called "best" design method. The relatively suitable design actually depends on the conditions. They are being summarized in Tables 4.11 and 4.12.

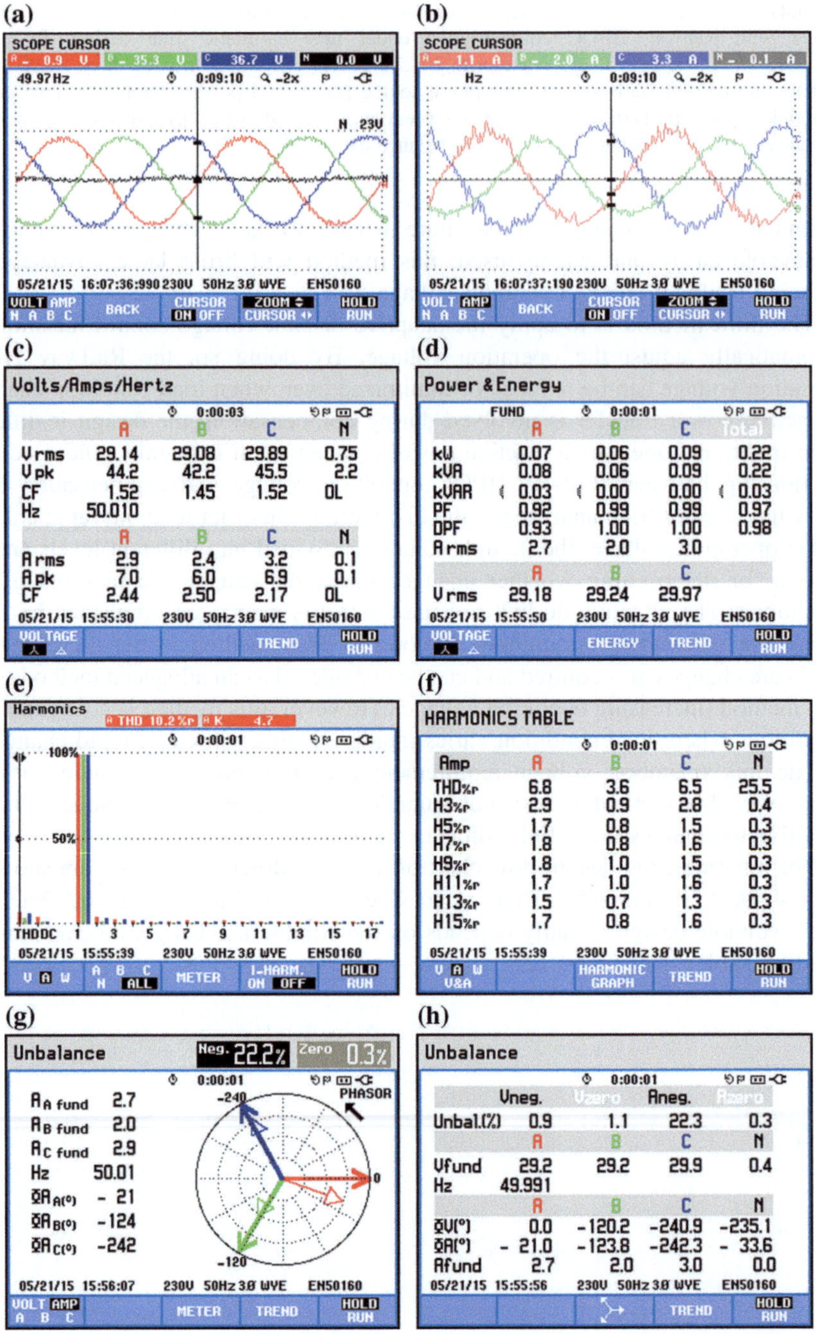
(a)
SCOPE CURSOR
49.97 Hz
0:09:10
N 23V
05/21/15 16:07:36:990 230V 50Hz 3Ø WYE EN50160
VOLT AMP
BACK
CURSOR ON OFF
ZOOM CURSOR
HOLD RUN
(b)
SCOPE CURSOR
Hz
0:09:10
05/21/15 16:07:37:190 230V 50Hz 3Ø WYE EN50160
VOLT AMP
BACK
CURSOR ON OFF
ZOOM CURSOR
HOLD RUN
(c)
Volts/Amps/Hertz
0:00:03
A B C N
Vrms 29.14 29.08 29.89 0.75
Vpk 44.2 42.2 45.5 2.2
CF 1.52 1.45 1.52 OL
Hz 50.010
A B C N
Arms 2.9 2.4 3.2 0.1
Apk 7.0 6.0 6.9 0.1
CF 2.44 2.50 2.17 OL
05/21/15 15:55:30 230V 50Hz 3Ø WYE EN50160
VOLTAGE
TREND
HOLD RUN
(d)
Power & Energy
FUND
0:00:01
A B C Total
kW 0.07 0.06 0.09 0.22
kVA 0.08 0.06 0.09 0.22
kVAR 0.03 0.00 0.00 0.03
PF 0.92 0.99 0.99 0.97
DPF 0.93 1.00 1.00 0.98
Arms 2.7 2.0 3.0
A B C
Vrms 29.18 29.24 29.97
05/21/15 15:55:50 230V 50Hz 3Ø WYE EN50160
VOLTAGE
ENERGY
TREND
HOLD RUN
(e)
Harmonics
THD 10.2 %r
K 4.7
0:00:01
100%
50%
THDDC 1 3 5 7 9 11 13 15 17
05/21/15 15:55:39 230V 50Hz 3Ø WYE EN50160
V A W
A B C N ALL
METER
I-HARM. ON OFF
HOLD RUN
(f)
HARMONICS TABLE
0:00:01
Amp A B C N
THD%r 6.8 3.6 6.5 25.5
H3%r 2.9 0.9 2.8 0.4
H5%r 1.7 0.8 1.5 0.3
H7%r 1.8 0.8 1.6 0.3
H9%r 1.8 1.0 1.5 0.3
H11%r 1.7 1.0 1.6 0.3
H13%r 1.5 0.7 1.3 0.3
H15%r 1.7 0.8 1.6 0.3
05/21/15 15:55:39 230V 50Hz 3Ø WYE EN50160
V A W V&A
HARMONIC GRAPH
TREND
HOLD RUN
(g)
Unbalance
Neg. 22.2%
Zero 0.3%
0:00:01
PHASOR
A A fund 2.7
A B fund 2.0
A C fund 2.9
Hz 50.01
ØA A(°) - 21
ØA B(°) -124
ØA C(°) -242
-240
0
-120
05/21/15 15:56:07 230V 50Hz 3Ø WYE EN50160
VOLT AMP A B C
METER
TREND
HOLD RUN
(h)
Unbalance
0:00:01
Vneg. Vzero Aneg. Azero
Unbal.(%) 0.9 1.1 22.3 0.3
A B C N
Vfund 29.2 29.2 29.9 0.4
Hz 49.991
A B C N
ØV(°) 0.0 -120.2 -240.9 -235.1
ØA(°) - 21.0 -123.8 -242.3 - 33.6
Afund 2.7 2.0 3.0 0.0
05/21/15 15:55:56 230V 50Hz 3Ø WYE EN50160
TREND
HOLD RUN

◀**Fig. 4.48** Experimental results of screens captured from hardware prototype of co-phase traction power with Railway HPQC compensation under new harmonic filter design (PF = 1.0, V_{dc} = 72 V): **a** source voltage waveform; **b** source current waveform; **c** three-phase source voltage and current statistics; **d** three-phase source power and power factor; **e** source current harmonics spectrum; **f** three-phase source current harmonics statistics; **g** source current negative sequence; **h** source voltage and current unbalance data

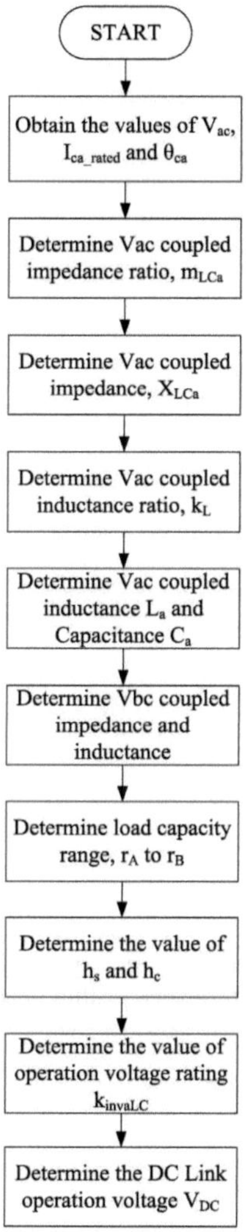

Important Equation

$$m_{LCa} = \sin\theta_{ca\_rated}$$

$$|X_{LCa}| = m_{LCa} \cdot \frac{V_{ac}}{I_{ca\_rated}} = \frac{V_{ac}\sin\theta_{ca}}{I_{ca\_rated}}$$

$$k_L = \frac{\sum_{h=2}^{\infty}\left(r_{h\_h}^2\right)\left[\frac{1}{h^2}\left(h^2-1\right)\right]}{\sum_{h=2}^{\infty}\left(r_{h\_h}^2\right)\left[\frac{1}{h^2}\left(h^2-1\right)^2\right]}$$

$$L_a = \frac{k_L|X_{LCa}|}{\omega} = \left(\frac{\sum_{h=2}^{\infty}\left(r_{h\_h}^2\right)\left[\frac{1}{h^2}\left(h^2-1\right)\right]}{\sum_{h=2}^{\infty}\left(r_{h\_h}^2\right)\left[\frac{1}{h^2}\left(h^2-1\right)^2\right]}\right)\cdot\left(\frac{|X_{LCa}|}{\omega}\right) \qquad C_a = \left(-1-\frac{\sum_{h=2}^{\infty}\left(r_{h\_h}^2\right)\left[\frac{1}{h^2}\left(h^2-1\right)\right]}{\sum_{h=2}^{\infty}\left(r_{h\_h}^2\right)\left[\frac{1}{h^2}\left(h^2-1\right)^2\right]}\right)\cdot\left(\frac{1}{\omega\cdot|X_{LCa}|}\right)$$

$$|X_{LCb}| \geq \frac{V_{bc}\sin\theta_{cb} - \sqrt{V_{invaLC\_min}^2 - V_{bc}^2\cos^2\theta_{cb}}}{I_{cb}} \qquad L_b = \frac{|X_{LCb}|}{\omega}$$

$$V_{DC} = \sqrt{2\left\{\left(1-m_{LCa}\sin\theta_{ca}\right)^2 + \left(m_{LCa}\cos\theta_{ca}\right)^2 + \sum_{h=2}^{\infty}\left(r_{h\_h}^2\right)\left[\frac{\left(\sin\theta_{ca}\right)^2}{A^2}\right]\left[-\frac{1}{h}\left[\left(h^2-1\right)k_L-1\right]\right]^2\right\}}$$

$$h_c = \frac{\cos\theta_{ca}}{\cos\theta_{ca\_rated}} \qquad h_s = \frac{\sin\theta_{ca}}{\sin\theta_{ca\_rated}}$$

$$k_{invaLC\_AB} \geq \sqrt{\left(1-r\cdot h_s\cdot\left(\sin\theta_{ca\_rated}\right)^2\right)^2 + \left(r\cdot h_c\cdot\sin\theta_{ca\_rated}\cdot\cos\theta_{ca\_rated}\right)^2}$$

$$V_{DC} = \sqrt{2\left(1-r\cdot h_s\cdot\left(\sin\theta_{ca\_rated}\right)^2\right)^2 + 2\left(r\cdot h_c\cdot\sin\theta_{ca\_rated}\cdot\cos\theta_{ca\_rated}\right)^2}$$

Fig. 4.49 Design flowchart of Railway HPQC in co-phase traction power with increased operation voltage to provide satisfactory compensation performance within load variation range

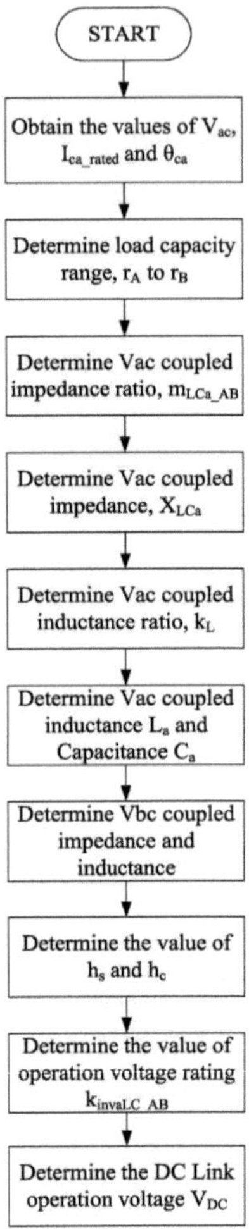

Important Equation

$$m_{LCa\_AB} = g \cdot m_{LCa\_rated} = \frac{2 \cdot (r_A h_{sA} - r_B h_{sB})}{(r_A + r_B)(r_A - r_B)} \cdot \sin\theta_{ca\_rated}$$

$$|X_{LCa}| = m_{LCa\_AB} \cdot \frac{V_{ac}}{I_{ca\_rated}} = \frac{2 \cdot (r_A h_{sA} - r_B h_{sB})}{(r_A + r_B)(r_A - r_B)} \cdot \sin\theta_{ca\_rated} \cdot \frac{V_{ac} \sin\theta_{ca\_rated}}{I_{ca\_rated}}$$

$$k_L = \frac{\sum_{h=2}^{\infty} \left(r_{h\_A}^2\right)\left[\frac{1}{h^2}(h^2-1)\right]}{\sum_{h=2}^{\infty} \left(r_{h\_A}^2\right)\left[\frac{1}{h^2}(h^2-1)^2\right]}$$

$$L_a = \frac{k_L |X_{LCa}|}{\omega} = \left(\frac{\sum_{h=2}^{\infty} \left(r_{h\_A}^2\right)\left[\frac{1}{h^2}(h^2-1)\right]}{\sum_{h=2}^{\infty} \left(r_{h\_A}^2\right)\left[\frac{1}{h^2}(h^2-1)^2\right]}\right) \cdot \left(\frac{|X_{LCa}|}{\omega}\right) \qquad C_a = \left(-1 - \frac{\sum_{h=2}^{\infty} \left(r_{h\_A}^2\right)\left[\frac{1}{h^2}(h^2-1)\right]}{\sum_{h=2}^{\infty} \left(r_{h\_A}^2\right)\left[\frac{1}{h^2}(h^2-1)^2\right]}\right) \cdot \left(\frac{1}{\omega \cdot |X_{LCa}|}\right)$$

$$|X_{LCb}| \geq \frac{V_{bc} \sin\theta_{cb} - \sqrt{V_{invaLC\_min}^2 - V_{bc}^2 \cos^2\theta_{cb}}}{I_{cb}} \qquad L_b = \frac{|X_{LCb}|}{\omega}$$

$$h_c = \frac{\cos\theta_{ca}}{\cos\theta_{ca\_rated}} \qquad h_s = \frac{\sin\theta_{ca}}{\sin\theta_{ca\_rated}}$$

$$k_{invaLC\_AB} = \sqrt{(r_A)^2 (m_{LCa\_AB})^2 - m_{LCa\_AB} \cdot (2 \cdot r_A \cdot h_{sA} \cdot \sin\theta_{ca\_rated}) + 1}$$
$$= \sqrt{(r_B)^2 (m_{LCa\_AB})^2 - m_{LCa\_AB} \cdot (2 \cdot r_B \cdot h_{sB} \cdot \sin\theta_{ca\_rated}) + 1}$$

$$V_{DC} = \sqrt{2(r_A)^2 (m_{LCa\_AB})^2 - 2m_{LCa\_AB} \cdot (2 \cdot r_A \cdot h_{sA} \cdot \sin\theta_{ca\_rated}) + 2}$$
$$= \sqrt{2(r_B)^2 (m_{LCa\_AB})^2 - 2m_{LCa\_AB} \cdot (2 \cdot r_B \cdot h_{sB} \cdot \sin\theta_{ca\_rated}) + 2}$$

Fig. 4.50 Design flowchart of Railway HPQC in co-phase traction power with impedance-mapping technique to provide satisfactory compensation performance within load variation range using lower operation voltage and coupled impedance

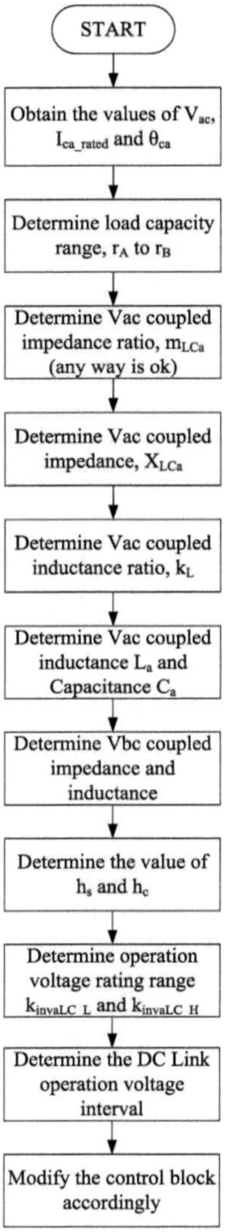

Important Equation

$$m_{pqL} = \left\{ \frac{2 \cdot \left[\frac{1}{2\sqrt{3}} (PF_{L\_L}) + \sqrt{1 - (PF_{L\_L})^2} \right]}{PF_{L\_L}} \right\}$$

$$m_{pqH} = \left\{ \frac{2 \cdot \left[\frac{1}{2\sqrt{3}} (PF_{L\_H}) + \sqrt{1 - (PF_{L\_H})^2} \right]}{PF_{L\_H}} \right\}$$

$$W = \sqrt{\left(1 - (m_{pqL})\left(\frac{r_L \cdot m_{LCa}}{\sqrt{1 + (m_{pqL})^2}}\right)\right)^2 + \frac{(r_L \cdot m_{LCa})^2}{\sqrt{1 + (m_{pqL})^2}}}$$

$$X = \sqrt{\left(1 - (m_{pqL})\left(\frac{r_H \cdot m_{LCa}}{\sqrt{1 + (m_{pqL})^2}}\right)\right)^2 + \frac{(r_H \cdot m_{LCa})^2}{\sqrt{1 + (m_{pqL})^2}}}$$

$$Y = \sqrt{\left(1 - (m_{pqH})\left(\frac{r_H \cdot m_{LCa}}{\sqrt{1 + (m_{pqH})^2}}\right)\right)^2 + \frac{(r_H \cdot m_{LCa})^2}{\sqrt{1 + (m_{pqH})^2}}}$$

$$Z = \sqrt{\left(1 - (m_{pqH})\left(\frac{r_L \cdot m_{LCa}}{\sqrt{1 + (m_{pqH})^2}}\right)\right)^2 + \frac{(r_L \cdot m_{LCa})^2}{\sqrt{1 + (m_{pqH})^2}}}$$

$$|X_{LCa}| = m_{LCa} \cdot \frac{V_{ac}}{I_{ca\_rated}}$$

$$k_L = \frac{\sum_{h=2}^{\infty} (r^2_{h\_h}) \left[\frac{1}{h^2}(h^2 - 1)\right]}{\sum_{h=2}^{\infty} (r^2_{h\_h}) \left[\frac{1}{h^2}(h^2 - 1)^2\right]}$$

$$L_a = \frac{k_L |X_{LCa}|}{\omega} = \left(\frac{\sum_{h=2}^{\infty} (r^2_{h\_h}) \left[\frac{1}{h^2}(h^2 - 1)\right]}{\sum_{h=2}^{\infty} (r^2_{h\_h}) \left[\frac{1}{h^2}(h^2 - 1)^2\right]} \right) \cdot \left(\frac{|X_{LCa}|}{\omega} \right) \qquad C_a = \left(-1 - \frac{\sum_{h=2}^{\infty} (r^2_{h\_h}) \left[\frac{1}{h^2}(h^2 - 1)\right]}{\sum_{h=2}^{\infty} (r^2_{h\_h}) \left[\frac{1}{h^2}(h^2 - 1)^2\right]} \right) \cdot \left(\frac{1}{\omega \cdot |X_{LCa}|} \right)$$

$$|X_{LCb}| \geq \frac{V_{bc} \sin\theta_{cb} - \sqrt{V^2_{invaLC\_min} - V^2_{bc} \cos^2\theta_{cb}}}{I_{cb}} \qquad L_b = \frac{|X_{LCb}|}{\omega}$$

$$h_c = \frac{\cos\theta_{ca}}{\cos\theta_{ca\_rated}} \qquad h_s = \frac{\sin\theta_{ca}}{\sin\theta_{ca\_rated}}$$

$$\begin{cases} k_{invaLC\_L} = \min(W, X, Y, Z) \\ k_{invaLC\_H} = \max(W, X, Y, Z) \end{cases}$$

$$V_{DC\_interval} = \sqrt{2} \cdot \left(\frac{k_{invaLC\_H} - k_{invaLC\_L}}{n} \right)$$

Fig. 4.51 Design flowchart of Railway HPQC in co-phase traction power with adaptive dc link voltage control technique to provide satisfactory compensation performance within load variation range with minimized operation voltage

Table 4.11 Summary showing the comparisons of the applied conditions between the three difference methods for Railway HPQC operation in co-phase traction power under load variation conditions

	Methods	Preferred applied conditions
A	Enhanced operation voltage to increase Railway HPQC compensation capability	When the system is running and hardware changes are not possible
B	Impedance-mapping technique	When system is being designed based on designed compensation range at initial stage
C	Adaptive dc link control technique	When lower dc link voltage can be used to reduce switching loss; and that control is not sensitive to dc voltage

Table 4.12 Summary showing the comparisons of pros and cons between the three different methods for Railway HPQC operation in co-phase traction power under load variation conditions

	Methods	Advantages	Disadvantages
A	Enhanced operation voltage to increase Railway HPQC compensation capability	Hardware parameters changes not required	Increased operation voltage may cause undesired power and switching loss under some conditions
B	Impedance-mapping technique	Can provide satisfactory compensation performance with lower dc operation voltage	Need changes in hardware parameters (coupled inductance and capacitance)
C	Adaptive dc link control technique	Can provide dynamic compensation performance over a range of loadings	Design of control parameters to reduce steady-state error and correct dc reference tracking is difficult; this degrades the compensation performance when the system control is highly dependent on dc voltage

References

1. Keng-Weng Lao; Man-Chung Wong; Ning Yi Dai; Chi-Kong Wong; Chi-Seng Lam, "Analysis of DC-Link Operation Voltage of a Hybrid Railway Power Quality Conditioner and Its PQ Compensation Capability in High-Speed Cophase Traction Power Supply", IEEE Transactions on Power Electronics, 2016, vol: 31, issue: 2, pp: 1643–1656.
2. Keng-Weng Lao; Man-Chung Wong; NingYi Dai; Chi-Seng Lam; Chi-Kong Wong; Lei Wang, "Analysis in the Effect of Co-phase Traction Railway HPQC Coupled Impedance on Its Compensation Capability and Impedance-Mapping Design Technique Based on Required Compensation Capability for Reduction in Operation Voltage", 2017, vol: 32, issue: 4, pp: 2631–2646.
3. Keng-Weng Lao; Man-Chung Wong; Ningyi Dai; Chi-Seng Lam; Lei Wang; Chi-Kong Wong, "Analysis of the Effects of Operation Voltage Range in Flexible DC Control on Railway HPQC Compensation Capability in High-Speed Co-phase Railway Power, " IEEE Transactions on Power Electronics, 2018, vol: 33, issue: 2, pp: 1760–1774.

Chapter 5
Partial Compensation Control in Co-phase Traction Power for Device Rating Reduction

Abstract Single-phase traction transformers are widely used in traction power supply systems due to their characteristics such as low cost and simple structure. Railway power conditioners (RPCs) could increase the loading capacity of the substations and improve the supply quality. In this paper, a Railway Hybrid Quality Conditioner (Railway HPQC) is applied to reduce reactive power, relieve system unbalance, and suppress harmonics. The operational voltage of the Railway HPQC is much lower than conventional RPC. A method named partial compensation was used to reduce current rating of the RPC, in which the compensating currents are expressed as a function of power factor target at the grid side. When this method is used in the HPRC, the operational voltage varies with the compensating currents over a wide range. A comprehensive design procedure is developed, so that both current and voltage ratings of the Railway HPQC are reduced with partial compensation. A reduction of more than 50% is achieved by compensating the power factor at the grid side to 0.95 instead of to unity. The control block diagram for the Railway HPQC with partial compensation is presented. The design and control of the Railway HPQC are validated by simulation and experimental results.

5.1 Introduction and Concept of Partial Compensation

The importance of railway electrification system in transportation systems let various countries put much effort and focus on electrified railway development. The alternating current (ac) is used for long-distance railway in countries such as United States, China, India, and France [1, 2]. Figure 5.1 shows a typical configuration of a single-phase 25 kV traction power system.

Tractions substations are used to transform power from three-phase power grid to single-phase ones and to supply power for locomotives in traction power system. The low cost and simple structure characteristics of single-phase transformer make it widely used in traction power [3–5]. As shown in Fig. 5.1, single-phase transformers are connected alternatively to the three phases in different sections to reduce unbalance currents. However, system unbalance can rarely be totally eliminated since there is hardly any occasion that locomotives are distributed evenly

K.-W. Lao et al., *Co-phase Traction Power Supply with Railway Hybrid Power Quality Conditioner*, https://doi.org/10.1007/978-981-13-0438-5_5

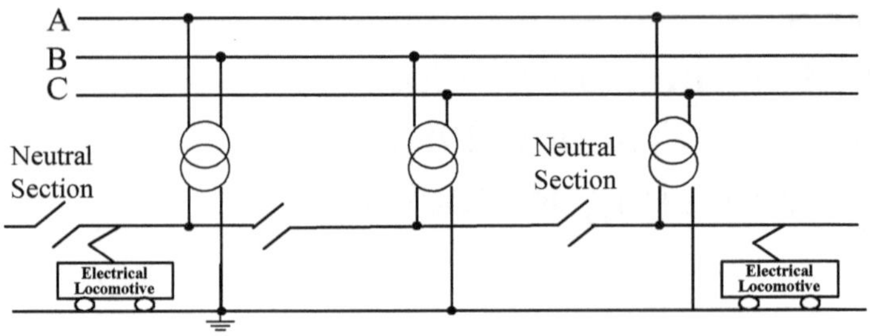

Fig. 5.1 Traction power supply system using single-phase traction transformers

along different sections [6, 7]. Furthermore, electric locomotives inject reactive power and harmonics to the traction power supply system, which cause the traction transformer to work in a derating mode and increase system losses [8–10].

Different power quality compensators have been proposed to solve the problems. For example, the Static VAR Compensator (SVC) was developed [11, 12]. However, SVC dynamic performance is poor and it occupies large area physically. Compared to SVC, the Active Power Filter (APC) and static synchronous compensator (STATCOM) have advantages such as fast response, high efficiency, and lower harmonic injections [13, 14]. The Railway Power quality Compensator (RPC) was also developed for traction power supply system, with traction transformers with two secondary windings, such as V/V transformer or Scott transformer [15, 16]. The RPC is composed a single-phase back-to-back converter to provide unbalance, reactive power, and harmonic current compensation at the same time [16–18]. The RPC was also used in a new traction power supply structure: co-phase traction power supply system. The first co-phase traction power device is now operating in Meishan Traction Substation in China.

One major drawback of the RPC and co-phase traction power supply is its high converter device rating and high cost. According to the field-recorded data [19], traction loads vary from 0 to around 40 MW. It is also investigated that during compensation, the RPC needs to transfer half of the load active power and all reactive power. Therefore, the rating of the RPC is high. In contrast, the Railway Hybrid Quality Conditioner (Railway HPQC) was proposed as a low-cost alternative to RPC in co-phase traction power supply system [20–22]. By having a capacitor in Railway HPQC coupling branch to make it capacitive, the operation voltage is significantly reduced compared to RPC. This leads to reduction in rating and operation loss of the Railway HPQC.

In order to satisfy the standard of power quality, traction substations in China are connected to 500 kV network to get sufficient short-circuit capacity [5]. At the same time, a number of existing traction substations are facing the challenges such as increased loading and deteriorated power quality. In this chapter, the Railway HPQC is applied to the traction power supply system in Fig. 5.1. The capacitive-coupling phase of the Railway HPQC is connected to the secondary side of the traction transformer. The other phase is connected to the high-voltage power grid. The power conditioning capability of the Railway HPQC can be used to provide compensation for reduction of reactive power, relieve system unbalance, and suppress harmonics.

In contrast to the co-phase traction power supply system, the conventional traction power supply systems allow a certain amount of unbalanced and/or distorted currents flowing in the system. Higher the short-circuit capacity of the system the more it is able to maintain its voltage in case of current distortion. Hence, the power conditioners are used to improve power factor or reduce current unbalance instead of fully solving the problems. Partial compensation is a more economical solution and it was introduced in [23] in which the compensating currents are expressed as a function of power factor at the grid side. The current rating of the RPC was effectively reduced with partial compensation, but its voltage rating is kept the same [23].

As discussed in Chap. 1, there is a tolerance standard for different power quality problems. In other words, the performance is considered satisfactory as long as the power quality is within tolerance. Partial compensation is applied to reduce both current and voltage ratings of the converter in the Railway HPQC [24]. The operational voltage of the Railway HPQC varies with the compensating currents over a wide range. Not only current rating, but also dc link voltage and coupling impedance need to be redesigned for the Railway HPQC under partial compensation. Details of the design and verifications are shown in the following sections.

5.2 System Model for Partial Compensation Investigation

This section shows the system model developed for partial compensation investigation. Based on its fundamental frequency model of Railway HPQC, a phasor diagram is shown in Fig. 5.2. The voltage at the high-voltage grid side is denoted as V_A, V_B, and V_C; while the secondary side voltage of the single-phase transformer is expressed as V_{ac} and V_{bc}. The main circuit of the Railway HPQC is a back-to-back converter, which absorbs power from the V_{bc} phase and injects power into the V_{ac} phase. The converter connecting to the V_{ac} phase is named the α-phase converter and the converter connecting to the V_{bc} phase is named the β-phase converter. The output currents of these two converters are denoted as $I_{c\alpha}$ and $I_{c\beta}$. The load current is I_L and its power factor is $cos(\varphi_L)$. The currents at the grid side are

$$\begin{bmatrix} \dot{I}_a \\ \dot{I}_b \\ \dot{I}_c \end{bmatrix} = \begin{bmatrix} \dot{I}_{\mathrm{aa}} \mathrm{e}^{-j\psi_a}/N_1 \\ \dot{I}_{\mathrm{bb}} \mathrm{e}^{-j\psi_b}/N_2 \\ -\dot{I}_a - \dot{I}_b \end{bmatrix} = \begin{bmatrix} (\dot{I}_L - \dot{I}_{c\alpha}) \mathrm{e}^{-j\psi_a}/N_1 \\ \dot{I}_{c\beta} \mathrm{e}^{-j\psi_b}/N_2 \\ -\dot{I}_a - \dot{I}_b \end{bmatrix} \tag{5.1}$$

where N_1 is the ratio of turns of the single-phase traction transformer, N_2 is the ratio of turns of coupling transformer at V_{bc} phase, $\psi_a = \pi/6$ and $\psi_b = \pi/2$.

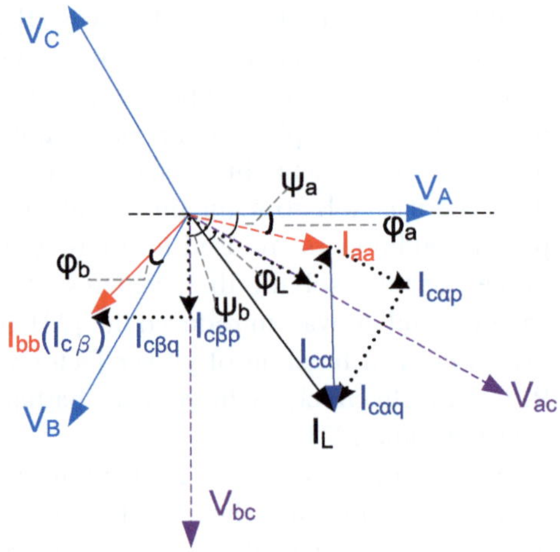

Fig. 5.2 Phasor diagram of the traction power supply system with Railway HPQC for partial compensation technique

5.3 Modified Control for Partial Compensation

5.3.1 Modified Control Function

According to the control of Railway HPQC for full compensation, the Railway HPQC transfers half load active power under full compensation. A parameter k is introduced to modify the percentage of load active power supplied by the Railway HPQC. It also affects reactive power being injected into the supply system because power factor is the ratio between active power and reactive power. Two more parameters, the phase angles φ_a and φ_b, are used to modify reactive power under partial compensation. They are obtained from power factor at the corresponding phase after the Railway HPQC operates. Hence, the reference currents are revised to (5.2) with partial compensation [23].

$$\begin{bmatrix} I_{c\alpha p} \\ I_{c\alpha q} \\ I_{c\beta p} \\ I_{c\beta q} \end{bmatrix} = \begin{bmatrix} k \cdot I_{\mathrm{Lp}} \\ \tan(\varphi_L) I_{\mathrm{Lp}} + \tan(\psi_a - \varphi_a)(1-k) I_{\mathrm{Lp}} \\ \frac{N_2}{N_1} \cdot k \cdot I_{\mathrm{Lp}} \\ \tan\left(\frac{2}{3}\pi - \psi_b + \varphi_b\right) \frac{N_2}{N_1} \cdot k \cdot I_{\mathrm{Lp}} \end{bmatrix} \tag{5.2}$$

5.3.2 Investigation on Current Ratings

Railway HPQC Current Rating for Full Compensation

The current rating for Railway HPQC under full compensation is presented in (5.3) and (5.4).

$$I_{c\alpha\_f} = \sqrt{\left(\frac{1}{2}\right)^2 + \left(\tan(\varphi_L) + \frac{\tan(\psi_a)}{2}\right)^2} I_{\mathrm{Lp}} \tag{5.3}$$

$$I_{c\beta\_f} = \sqrt{\left(\frac{1}{2}\right)^2 + \left(\frac{\tan\left(\frac{2}{3}\pi - \psi_b\right)}{2}\right)^2} \cdot \frac{N_2}{N_1} \cdot I_{\mathrm{Lp}} \tag{5.4}$$

Railway HPQC Current Rating for Partial Compensation

The current ratings of the two converters in the Railway HPQC with partial compensation are expressed as follows.

$$I_{c\alpha} = \sqrt{k^2 + (\tan(\varphi_L) + \tan(\psi_a - \varphi_a)(1-k))^2} I_{\mathrm{Lp}} \tag{5.5}$$

$$I_{c\beta} = \sqrt{(k)^2 + \left(\tan\left(\frac{2}{3}\pi - \psi_b + \varphi_b\right) \cdot k\right)^2} \cdot \frac{N_2}{N_1} \cdot I_{\mathrm{Lp}} \tag{5.6}$$

It is assumed that load power factor is 0.85. The current ratings variation in terms of power factor and k are shown in Fig. 5.3. The ratings in Fig. 5.3 are normalized by dividing the rating by that required for full compensation. It is concluded from Fig. 5.4 that current rating of the α-phase converter does not vary linearly with k. There is one minimum point for each selected power factor target at phase A. The current rating of the β-phase converter varies linearly with k. When the power factor is lower than 1 and an appropriate value is selected for k, the current rating reduction is achieved with partial compensation.

5.3.3 Voltage Ratings with Partial Compensation

In this part, the voltage ratings of power converters in the Railway HPQC are calculated when partial compensation is applied. The output voltage of the α-phase converter and its rating are expressed as follows.

$$\vec{V}_{\mathrm{inv}\alpha} = \vec{V}_{\mathrm{ac}} + \vec{V}_{\mathrm{LC}} = \vec{V}_{\mathrm{ac}} - jX_{\mathrm{LC}} \cdot \vec{i}_{c\alpha} \tag{5.7}$$

$$V_{\mathrm{inv}\alpha} = \sqrt{\left(V_{\mathrm{ac}} - I_{c\alpha q}X_{\mathrm{LC}}\right)^2 + \left(I_{c\alpha p}X_{\mathrm{LC}}\right)^2} \tag{5.8}$$

For a fixed compensating current, the optimum parameter selection of the coupling impedance X_{LC} is achieved by taking the derivative of (5.8) with X_{LC} and setting it to zero. The process and results are shown in (5.9) and (5.10).

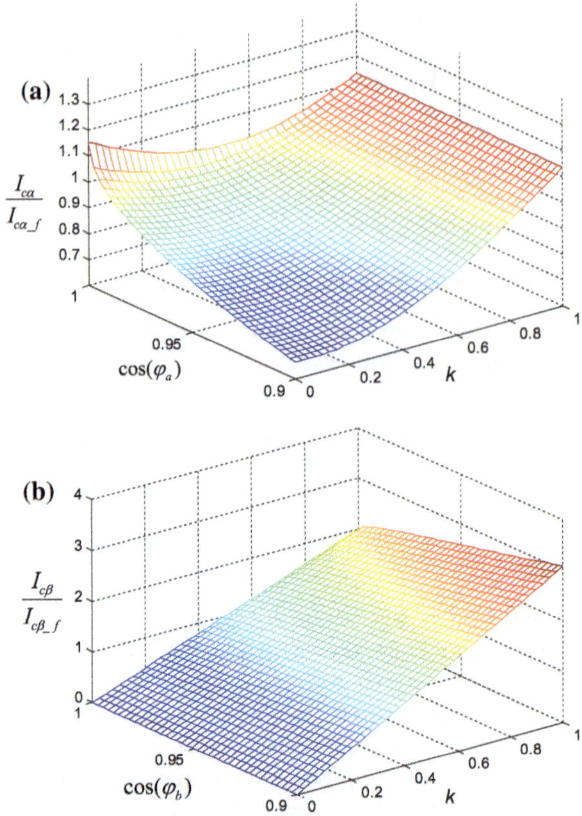

Fig. 5.3 Current ratings **a** α-phase converter **b** β-phase converter

$$\frac{\mathrm{d}(V_{\mathrm{inv}\alpha}^2)}{\mathrm{d}(X_{\mathrm{LC}})} = 2V_{\mathrm{ac}}I_{c\alpha q} + 2\left(I_{c\alpha q}^2 + I_{c\alpha p}^2\right)X_{\mathrm{LC}} = 0 \tag{5.9}$$

$$X_{\mathrm{LC}} = \frac{\tan(\varphi_L) + \tan(\psi_a - \varphi_a)(1-k)}{\left((\tan(\varphi_L) + \tan(\psi_a - \varphi_a)(1-k)\right)^2 + k^2} \cdot \frac{V_{\mathrm{ac}}}{I_{\mathrm{Lp}}} \tag{5.10}$$

By substituting (5.3) and (5.10) into (5.8), the voltage rating of the α-phase converter is deduced as

$$V_{\mathrm{inv}\alpha} = \frac{k}{\sqrt{\left(\tan(\varphi_L) + \tan(\psi_a - \varphi_a)(1-k)\right)^2 + k^2}} \cdot V_{\mathrm{ac}} \tag{5.11}$$

The voltage rating with full compensation is given as

$$V_{\mathrm{inv}\alpha\_f} = \frac{1/2}{\sqrt{(\tan(\varphi_L) + \tan(\psi_a)(1/2))^2 + (1/2)^2}} \cdot V_{\mathrm{ac}} \quad (5.12)$$

The output voltage of the β-phase converter and its rating are expressed as follows.

$$\vec{V}_{\mathrm{inv}\beta} = \vec{V}_{\mathrm{bc}} - \vec{V}_L = \vec{V}_{\mathrm{bc}} - jX_L \cdot \vec{i}_{c\beta} \quad (5.13)$$

$$V_{\mathrm{inv}\beta} = \sqrt{\left(V_{\mathrm{bc}} - I_{c\beta q}X_L\right)^2 + \left(I_{c\beta p}X_L\right)^2} \quad (5.14)$$

The coupling inductor of the β-phase converter is selected to suppress the output current ripple and is kept as small as possible. Its value is assumed to be

$$X_L = m \cdot \frac{V_{\mathrm{bc}}}{(N_2/N_1)I_{\mathrm{Lp}}} \quad (5.15)$$

By substituting (5.2) and (5.15) into (5.14), the voltage rating of the β-phase converter is given in (5.16) and (5.17) with partial compensation and full compensation respectively.

$$V_{\mathrm{inv}\beta} = \sqrt{\left(1 - \tan\left(\frac{2}{3}\pi - \psi_b + \varphi_b\right) \cdot m \cdot k\right)^2 + k^2 \cdot m^2} \cdot V_{\mathrm{bc}} \quad (5.16)$$

$$V_{\mathrm{inv}\beta\_f} = \sqrt{\left(1 - \frac{1}{2}\tan\left(\frac{2}{3}\pi - \psi_b\right) \cdot m\right)^2 + \frac{1}{4} \cdot m^2} \cdot V_{\mathrm{bc}} \quad (5.17)$$

The variations of voltage rating in terms of the power factor and k are illustrated in Fig. 5.4. The voltage rating of the α-phase converter varies over a wide range with partial compensation. The coefficient m is set to 10% when voltage ratings are calculated in Fig. 5.4b. In contrast to the voltage rating of the α-phase converter, the voltage rating of the β-phase converter varies over a narrower range. The reason is that variation in voltage crossing the coupling impedance is higher at the α-phase than at the β-phase. When the power factor is lower than 1 and an appropriate value is selected for k, the voltage rating reduction is also achieved with partial compensation. In summary, the voltage rating needs to be adjusted together with current rating in Railway HPQC. But it is not the case in RPC when partial compensation is implemented.

5.3.4 Railway HPQC Rating Under Partial Compensation

In this part, the rating of the power converters in the Railway HPQC is calculated as follows.

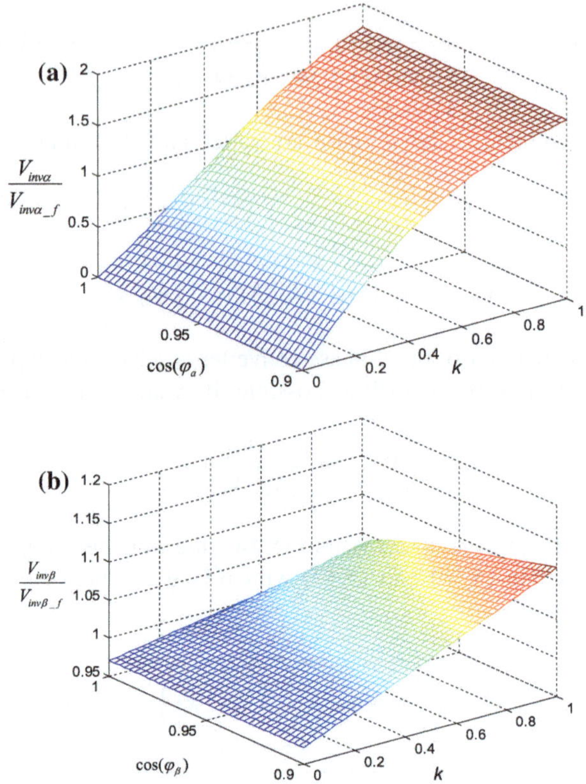

Fig. 5.4 Voltage ratings **a** α-phase converter **b** β-phase converter

$$S_{\text{HPQC}} = S_\alpha + S_\beta = V_{\text{inv}\alpha} \cdot I_{c\alpha} + V_{\text{inv}\beta} \cdot I_{c\beta} \tag{5.18}$$

It is assumed that cos (φ_a) = cos (φ_b). The variation of the converter rating in terms of the power factor and k is shown in Fig. 5.5.

5.4 Railway HPQC Design with Partial Compensation

5.4.1 Parameter Selection for Partial Compensation

As illustrated in Fig. 5.5, designing with partial compensation can significantly reduce the rating of the power converters. However, power factor at the grid side is not able to reach unity under partial compensation. The variation of compensating current also affects the performance of current balancing. There is a trade-off between the Railway HPQC rating reduction and power conditioning performance.

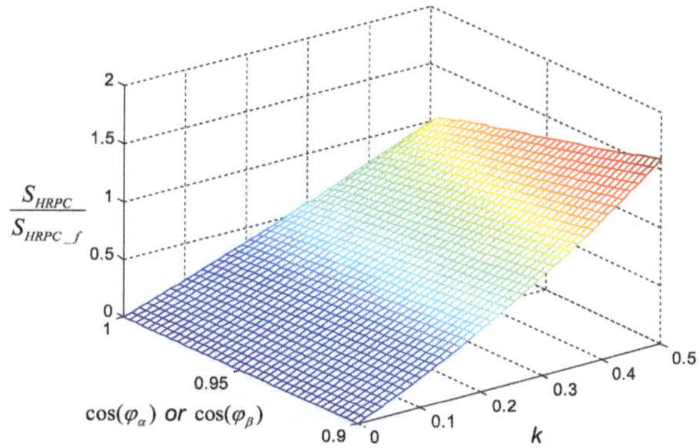

Fig. 5.5 Variation of the converter rating in terms of the power factor and k

There are three parameters affecting compensating currents in (5.2): k, $\varphi_{a,}$ and φ_b. The last two are determined by the grid-side power factor. There is a relationship between the parameter k and grid-side power factor, as given in (5.19) [23].

$$k = \frac{\cos\left(\psi_b - \varphi_b - \frac{2}{3}\pi\right)\sin\left(\varphi_a - \varphi_c + \frac{2}{3}\pi\right)}{\cos\left(\psi_b - \varphi_b - \frac{2}{3}\pi\right)\sin\left(\varphi_a - \varphi_c + \frac{2}{3}\pi\right) + \cos\left(\psi_a - \varphi_a\right)\sin\left(\varphi_c - \varphi_b + \frac{2}{3}\pi\right)} \tag{5.19}$$

To reach the same grid-side power factor, the phase angle of each phase could be either leading or lagging. The k value varies in terms of the phase angle for each phase. Previous study verified that the minimum current rating is obtained by setting the power angle of phases A and B as lagging and the power angle of phase C as leading under partial compensation [23] This scheme is also used in this paper to design the Railway HPQC with partial compensation.

The effects of k value on grid-side power factor and Railway HPQC rating are shown in Fig. 5.6. The Railway HPQC rating decreases almost linearly with k. The k decreases much faster when the power factor is close to unity. The Railway HPQC rating has a reduction of about 20% by setting the power factor to 0.995 instead of 1. As a result, the partial compensation can achieve Railway HPQC rating reduction without sacrificing the system performance too much.

5.4.2 Comprehensive Design Procedure of the Railway HPQC Under Partial Compensation

With the selected power factor and k, the coupling impedance and dc link voltage of the Railway HPQC needs to be adjusted accordingly. The comprehensive design procedure of the Railway HPQC is presented as follows.

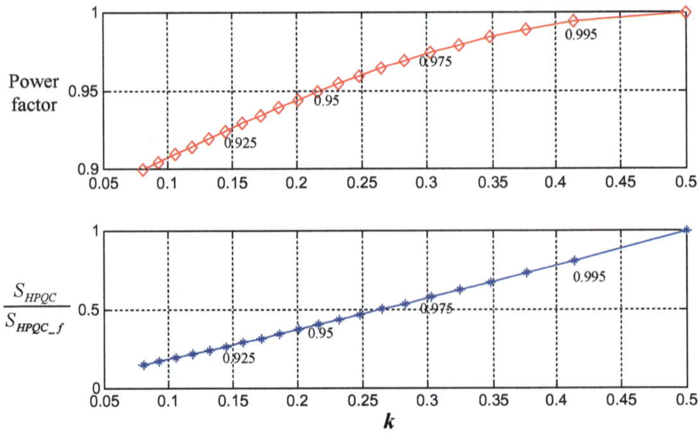

Fig. 5.6 Effect of k on power factor and railway HPQC rating

1. Set the power factor target at the grid side after the Railway HPQC operates.

$$PF_{\text{grid}} = \frac{(P_A + P_B + P_C)}{(S_A + S_B + S_C)} \tag{5.20}$$

2. The partial compensation coefficient k is calculated by

$$k = \frac{\cos\left(\psi_b - \varphi_b - \frac{2}{3}\pi\right)\sin\left(\varphi_a - \varphi_c + \frac{2}{3}\pi\right)}{\cos\left(\psi_b - \varphi_b - \frac{2}{3}\pi\right)\sin\left(\varphi_a - \varphi_c + \frac{2}{3}\pi\right) + \cos(\psi_a - \varphi_a)\sin\left(\varphi_c - \varphi_b + \frac{2}{3}\pi\right)} \tag{5.21}$$

$$\varphi_a = \varphi_b = -\varphi_c = a\cos(PF_{\text{grid}}) \tag{5.22}$$

3. The reference currents for controlling the Railway HPQC are given in (5.23), in which $k_\alpha = \tan(\psi_a - \varphi_a)(1 - k)$ and $k_\beta = \tan\left(\frac{2}{3}\pi - \psi_b + \varphi_b\right)$.

$$\begin{bmatrix} I_{c\alpha p} \\ I_{c\alpha q} \\ I_{c\beta p} \\ I_{c\beta q} \end{bmatrix} = \begin{bmatrix} k \cdot I_{\text{Lp}} \\ \tan(\varphi_L) I_{\text{Lp}} + \tan(\psi_a - \varphi_a)(1 - k) I_{\text{Lp}} \\ \frac{N_2}{N_1} \cdot k \cdot I_{\text{Lp}} \\ \tan\left(\frac{2}{3}\pi - \psi_b + \varphi_b\right) \frac{N_2}{N_1} \cdot k \cdot I_{\text{Lp}} \end{bmatrix} = \begin{bmatrix} k \cdot I_{\text{Lp}} \\ \tan(\varphi_L) I_{\text{Lp}} + k_\alpha I_{\text{Lp}} \\ \frac{N_2}{N_1} \cdot k \cdot I_{\text{Lp}} \\ \frac{N_2}{N_1} \cdot k_\beta \cdot k \cdot I_{\text{Lp}} \end{bmatrix} \tag{5.23}$$

4. Calculate the coupling impedance of the α-phase converter.

$$X_{\mathrm{LC}} = \frac{\tan(\varphi_L) + \tan(\psi_a - \varphi_a)(1-k)}{((\tan(\varphi_L) + \tan(\psi_a - \varphi_a)(1-k))^2 + k^2} \cdot \frac{V_{\mathrm{ac}}}{I_{\mathrm{Lp}}} \tag{5.24}$$

5. Calculate the α-phase coupling inductor and capacitor taking the harmonic compensation into consideration. The load harmonic current at the hth harmonic is assumed to be r_h times the fundamental value, as given in (5.25). The coupling impedance is selected to minimize the harmonic operational voltage, as given in (5.26) and (5.27) [22].

$$I_{\mathrm{Lh}} = r_h \cdot I_{L1} = r_h \cdot I_{L1p} / \cos(\varphi_L) \tag{5.25}$$

$$L_a = \frac{\dfrac{\sum_{h=2}^{\infty}(r_h)^2 \cdot \frac{2(h^2-1)}{h^2}}{\sum_{h=2}^{\infty}(r_h)^2 \cdot \frac{2(h^2-1)^2}{h^2}}}{\omega_1} X_{\mathrm{LC}} = \frac{k_L}{\omega_1} X_{\mathrm{LC}} \tag{5.26}$$

$$C_a = \frac{1}{\omega_1\left(1 + \dfrac{\sum_{h=2}^{\infty}(r_h)^2 \cdot \frac{2(h^2-1)}{h^2}}{\sum_{h=2}^{\infty}(r_h)^2 \cdot \frac{2(h^2-1)^2}{h^2}}\right) \cdot X_{\mathrm{LCa}}} = \frac{1}{\omega_1 k_c \cdot X_{\mathrm{LC}}} \tag{5.27}$$

6. Calculate the β-phase coupling inductor according to (5.15) and (5.28).

$$L_\beta = \frac{X_L}{\omega_1} \tag{5.28}$$

7. Determine the dc link operation voltage in the Railway HPQC as follows.

$$V_{\mathrm{dc}} = \sqrt{2}\sqrt{V_{\mathrm{inv}\alpha}^2 + \sum_{h=2}^{\infty}(X_{\mathrm{LCh}} \cdot I_{\mathrm{Lh}})^2} \tag{5.29}$$

Where $X_{\mathrm{LCh}} = -\frac{1}{h}[(h^2-1)k_L - 1]X_{\mathrm{LC}}$.

8. The α-phase and β-phase converter are connected to the same dc bus. But the operational voltage at the β-phase is much higher than that at the α-phase. The problem is solved by adjusting the ratio of turns of the coupling transformer at the β-phase. It is expressed as follows.

$$N_2 = \frac{V_{\mathrm{BC}}}{V_{\mathrm{dc}}/\sqrt{2}} \tag{5.30}$$

5.5 Modified Control System of Railway HPQC for Partial Compensation

Figure 5.7 illustrates the control diagram of the Railway HPQC.

The single-phase instantaneous theory is used to calculate the power of the traction loads. The instantaneous active and reactive powers are calculated as follows, in which $v_{\alpha d}$ and $i_{\alpha d}$ are the 90° delay of the system voltage and load current, respectively.

$$\begin{bmatrix} p \\ q \end{bmatrix} = \begin{bmatrix} v_\alpha \cdot i_L + v_{\alpha d} \cdot i_{Ld} \\ v_\alpha \cdot i_{Ld} - v_{\alpha d} \cdot i_L \end{bmatrix} \tag{5.31}$$

Instead of calculating the compensating current by using (5.23), partial compensation is achieved by multiplying the corresponding coefficient to the power reference, as given in (5.32).

$$\begin{bmatrix} p_{p\alpha} \\ q_{pa} \\ p_{p\beta} \\ q_{p\beta} \end{bmatrix} = \begin{bmatrix} k \cdot p_{\rm dc} + p_{\rm ac} \\ k_\alpha \cdot p_{\rm dc} + q \\ k \cdot p_{\rm dc} + p^* \\ k_\beta \cdot (p_{\rm dc} \cdot k + p^*) \end{bmatrix} \tag{5.32}$$

By varying these coefficients, the Railway HPQC is able to improve the power factor on the grid side in relation to the selected target. The reference current for controlling the converter is extracted from the power by inversing the transform in (5.31). The last stage of the control system is the PI current controller and Pulse Width Modulation (PWM) unit.

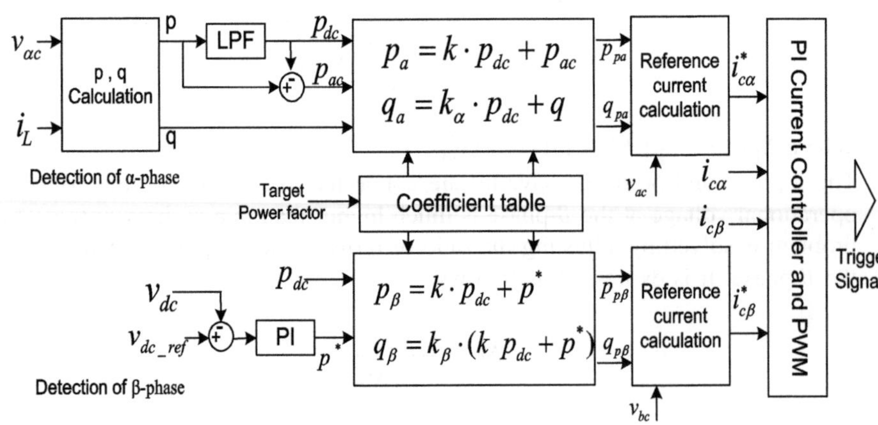

Fig. 5.7 Control system of the railway HPQC

5.6 Case Study and Simulation

In this section, a case study is presented to verify the analysis and the comprehensive design procedure developed in the previous parts. The circuit schematics can be found in Fig. 5.8. The traction substation uses one single-phase traction transformer. Its primary side is connected to a 110 kV power network and its secondary side provides a 27.5 kV supply to the traction loads. The traction load is 15 MVA, with a load power factor of 0.85. The system source impedance is calculated as 2 mH according to the short-circuit capacity of the common traction power supply of 750 MVA.

Simulation models are built using PSCAD/EMTDC. Table 5.1 shows the practical onsite data for the harmonic distribution in traction load for the WuQing substation in China [25]. The load harmonics are designed according to the data in Table 5.1. The grid-side voltages and currents without Railway HPQC are shown in

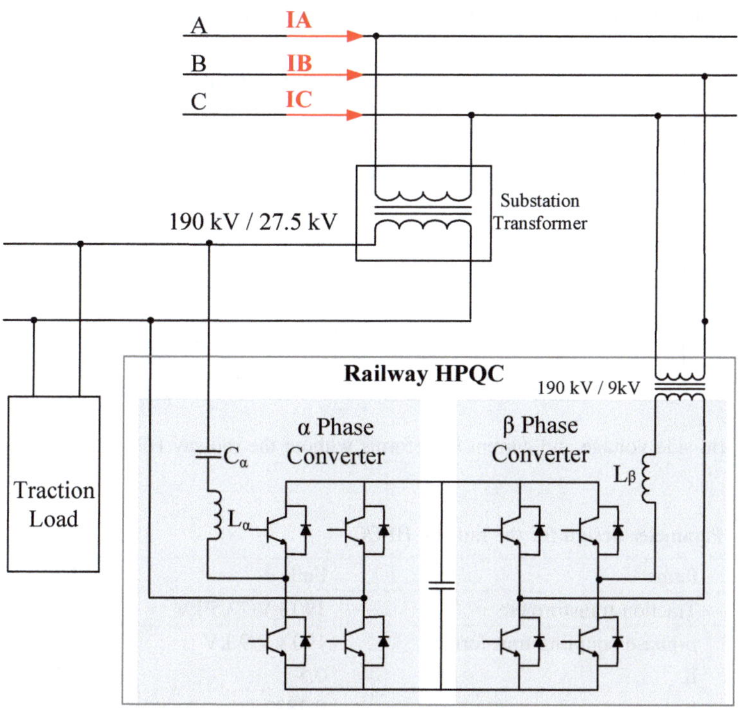

Fig. 5.8 System configuration for simulation case study

Table 5.1 On-site statistics for the harmonic current contents in The WuQing substation traction load

	3rd	5th	7th	9th	11th
Harmonic contents (% of fundamental)	10.81	7.96	4.51	3.04	2.68

Fig. 5.10, which shows that the traction load currents are severely unbalanced with a low power factor and high harmonics contents (Fig. 5.9).

A Railway HPQC is installed to improve the power quality of the WuQing substation in simulation. The parameters designed for the Railway HPQC with full compensation are listed in Table 5.2. Full compensation is first achieved by the Railway HPQC using a dc link voltage of 18.7 kV. The grid-side voltages and currents after compensation are shown in Fig. 5.10.

The three-phase currents are balanced with the unity power factor. The harmonics are also eliminated. The system performance is summarized in Table 5.3, in which the Railway HPQC rating is calculated by

$$S_{\text{HRPC}} = S_\alpha + S_\beta = \frac{V_{\text{dc}}}{\sqrt{2}}(I_{c\alpha} + I_{c\beta}) \tag{5.33}$$

The target for the partial compensation is set according to the utility tariff plan for reactive power. For example, the PF must be at least 0.9 to avoid a penalty [26]. For the traction motor, more reactive power is consumed during the motor starting.

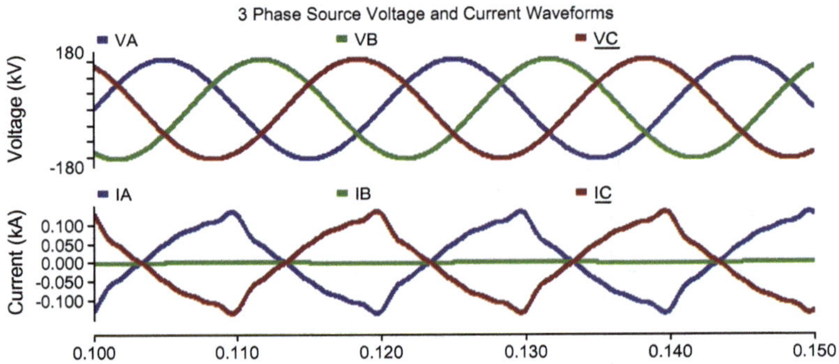

Fig. 5.9 Grid-side voltage and current waveforms without the railway HPQC

Table 5.2 Parameter design for the railway HPQC

No.	Items	Full	Partial
1	Traction transformer	190 kV/27.5 kV	
2	β-phase coupling transformer	190 kV/9 kV	
3	K	0.5	0.2154
	k_α	0.2887	0.1640
	k_β	0.5774	1.1182
4	α-phase coupling inductor L_α	6.6 mH	9.3 mH
5	α-phase coupling capacitor C_α	61.00 μF	43.45 μF
6	β-phase coupling inductor L_β	8 mH	10 mH
7	V_{dc}	18.7 kV	11 kV
8	Sampling frequency	20 kHz	

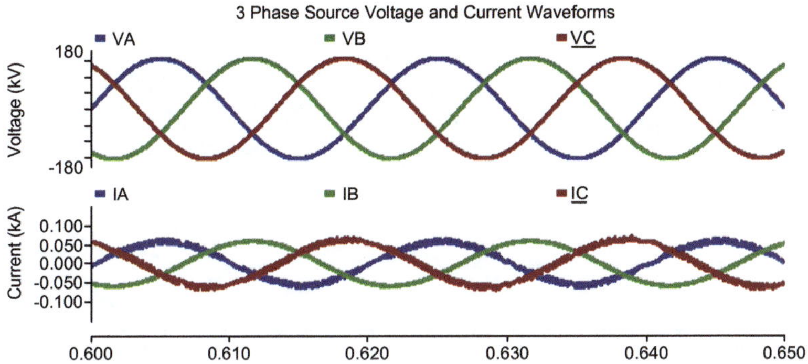

Fig. 5.10 Grid-side voltage and current waveforms with the railway HPQC settings for achieving full compensation (V_{dc} = 18.7 kV)

Table 5.3 System performance in simulation

Condition	Power factor	Source current THD (phase A) (%)	Current unbalance (%)	Voltage unbalance (%)	Railway HPQC rating (MVA)
Before compensation	0.60	14.7	100	2.59	–
Railway HPQC with full compensation (V_{dc} = 18.7 kV)	0.997	2.34	4.75	0.40	15.13
Railway HPQC with full compensation (V_{dc} = 11 kV)	0.985	37.9	22.63	0.84	10.5
Railway HPQC with partial compensation (V_{dc} = 11 kV)	0.954	3.27	44	1.03	6.97

It is better to over-design the reactive power compensation capability. A safety margin is required to take into account possible over-loading. In this chapter, 0.95 is used as the power factor to design the Railway HPQC. The system parameters with partial compensation are listed in Table 5.2.

By designing the Railway HPQC with partial compensation, the dc link voltage is reduced to 11 kV. The Railway HPQC is not able to achieve full compensation under this dc link voltage. Unsatisfactory grid-side current waveforms are shown in Fig. 5.11. It is obvious that the Railway HPQC fails to improve the power quality since the reduced dc link voltage is not high enough for the Railway HPQC to achieve full compensation. The control block diagram shown in Fig. 5.8 is adopted to modify the reference of the Railway HPQC to achieve partial compensation. The corresponding simulation results are illustrated in Fig. 5.12 and summarized in Table 5.3. The power factor is improved to around 0.95, the harmonics are suppressed and the current unbalance is reduced. The voltage unbalance is calculated

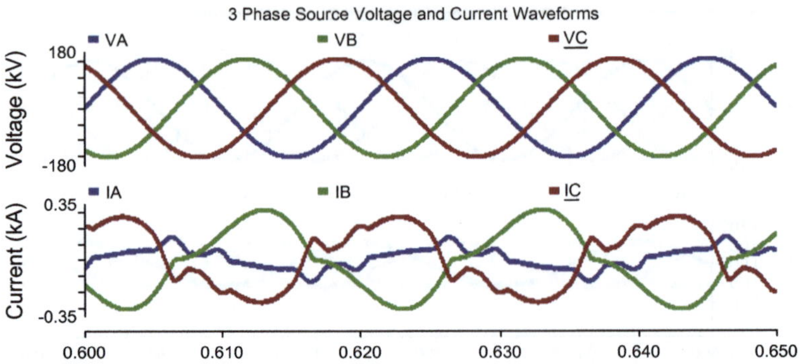

Fig. 5.11 Unsatisfactory grid-side voltage and current waveforms under reduced voltage with the Railway HPQC settings to achieve full compensation (V_{dc} = 11 kV)

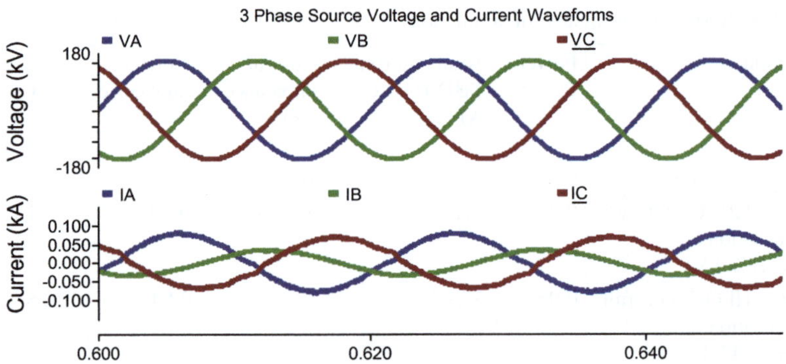

Fig. 5.12 Grid-side voltage and current waveforms with the Railway HPQC settings to achieve partial compensation (power factor = 0.95, V_{dc} = 11 kV)

and given in Table 5.3. Results indicate that voltage unbalance does not exceed 3% and satisfies the power quality standard [27]. Hence, the current unbalance is compensated to an acceptable level. The total rating of the Railway HPQC is significantly reduced. It is less than 50% of the rating with full compensation.

5.7 Experimental Results

A small-capacity Railway HPQC prototype was built with the same system configuration as given in Fig. 5.8. The peak value of the sinusoidal three-phase supply voltage is 70.7 V. The load comprises a single-phase rectifier, and its harmonic distribution is given in Table 5.4. The corresponding parameters for the experiment are listed in Table 5.5. The system current and voltage waveforms are monitored

Table 5.4 Harmonic current contents in the testing load

	3rd	5th	7th	9th	11th
Harmonic contents (% of fundamental)	27.9	5.7	2.0	1.5	1.2

Table 5.5 Parameter designed for railway HPQC in the Experiment

No.	Items	Full	Partial
1	Traction transformer	220 V/50 V	
2	β-phase coupling transformer	220 V/25 V	
3	α-phase coupling inductor L_α	6.7 mH	6.7 mH
4	α-phase coupling capacitor C_α	190 μF	120 μF
5	β-phase coupling inductor L_β	10 mH	10 mH
6	V_{dc}	41 V	24 V
7	Sampling frequency	20 kHz	

using a Yokogawa DL750 16-channel ScopeCorder Oscilloscope and the power quality is recorded using a Fluke 43B Power Quality Analyzer.

The voltage and current waveforms at the grid side are shown in Fig. 5.13a when only single-phase transform supplies the load. The Railway HPQC is first controlled to do full compensation. The system waveforms with the Railway HPQC under full compensation are presented in Fig. 5.13b. It can be concluded from Fig. 5.13a, b that current unbalance, reactive power, and harmonics are compensated simultaneously. The system performance is summarized in Table 5.6. The dc link voltage under this case is 41 V, which is lower than the grid-side voltage.

The Railway HPQC is tested by adjusting the compensation target. The power factor is set to 0.95 and the parameters in the control blocks are modified accordingly. The waveforms with the Railway HPQC under partial compensation are presented in Fig. 5.13c. The dc link voltage is reduced to 24 V. The reactive current and unbalance are partially compensated. The harmonics are eliminated illustrated in Fig. 5.13 and Table 5.6. With sufficient short-circuit capacity, the power quality standard is satisfied. The converter rating reduction of more than 50% is achieved with partial compensation. Not only is the initial cost of the system reduced, the operational losses are also reduced as the dc link operating voltage is lower.

5.8 Summary

Partial compensation is implemented to reduce the compensating current and operational voltage of the Railway HPQC. A comprehensive design procedure is designed for Railway HPQC to reduce converter rating with partial compensation. With new design method, the rating of the Railway HPQC is expressed as a

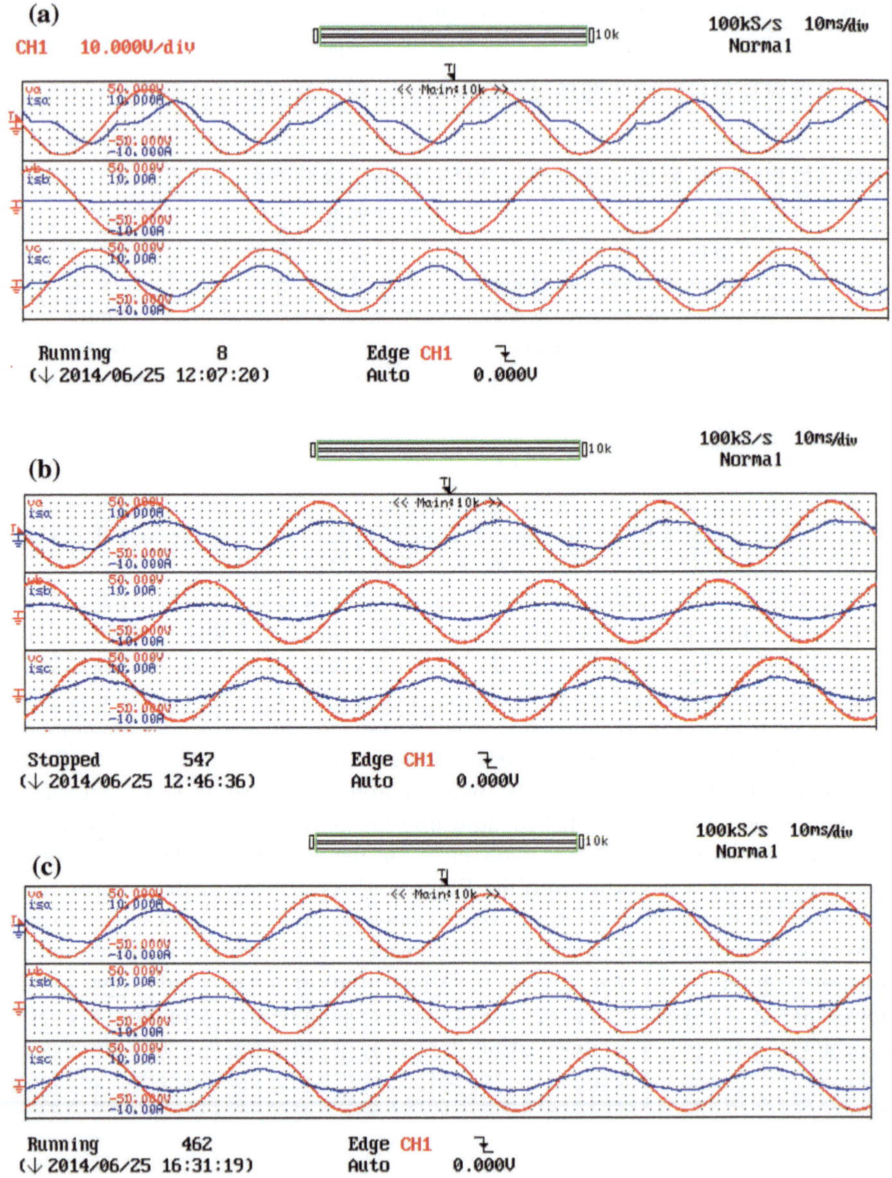

Fig. 5.13 Experimental results: three-phase voltage and current at the grid side **a** without railway HPQC **b** railway HPQC with full compensation **c** railway HPQC with partial compensation

function of the power factor at the grid side. For example, the Railway HPQC rating decreases by50% if the grid-side power factor is set to 0.95 instead of unity. The initial cost and size of the system are greatly reduced. The control system of Railway HPQC with partial compensation is implemented. The design and control

Table 5.6 System performance in the experiment

Condition	Power factor	Source current THD (phase A) (%)	Current unbalance (%)	Railway HPQC rating
Before compensation	0.76	28.6	92.2	–
Railway HPQC with full compensation	0.95	9.1	26	142.3 VA
Railway HPQC with partial compensation	0.94	6.9	43.3	62.97 VA

of Railway HPQC with partial compensation are verified by simulation. A small-capacity experimental prototype is built in the laboratory and testing results are also provided to verify the design and control method.

References

1. Y. Jiang, J. Q. Liu, W. Tian, M. Shahidehpour, and M. Krishamurthy, "Energy harvesting for the electrification of railway stations," IEEE Electrification Magazine, Sept. 2014, pp. 39–48.
2. A. Gomez-Exposito, J. M. Mauricio, and J. M. Maza-Ortega, "VSC-based MVDC railway electrification system," IEEE Trans. Power Delivery, Vol. 29, No. 1, Feb. 2014, pp. 422–431.
3. P-C. Tan, P.C. Loh, and D. G. Holmes, "Optimal impedance termination of 25-kV electrified railway systems for improved power quality," IEEE Trans. Power Delivery, Vol. 20, No. 2, pp. 1703–1710, Apr. 2005.
4. S. T. Senini, and P. J. Wolfs, "Novel topology for correction of unbalanced load in single phase electric traction systems," in PESC Rec. IEEE Annu. Power Electronics Specialists Conf., Vol. 3, Cairns, Australia, June 2002, pp. 1208–1212.
5. Y. Q. Liu, H. S. Hua, G. P. Wu, and L. Wang, "Research for the effects of high-speed electrified railway traction load on power quality," in Proc. of DRPT 2011, pp. 569–573, 2011.
6. H. Wang, Y. Tian, and Q. Gui, "Evaluation of negative sequence current injection into the public grid from different traction substation in electrical railways," in Proc. of the International Conference and Exhibition on Electricity Distribution, CIRED, 2009, pp. 1–4.
7. B. Fukala, and J. Palecek, "Comparision of schemes of traction transformer stations in terms of their impact on the asymmetry in the power supply system," in Proc. of 15th Intl. Conf. on Electric Power Engineering (EPE), 2014, pp. 207–210.
8. G. Lin, X. Yonghai, X. Xiangnin, et al., "Analysis of adverse effects on the public power grid brought by traction power-supply system," in Proc. 2008 IEEE Electrical Power and Energy Conf,.
9. T. Tanaka, K. Ishibashi, N. Ishikura, and E. Hiraki, "A half-bridge inverter based active power quality compensator for electrified railway," in Proc. of 2010 International Power Electronics Conf., 2010, pp. 1590–1595.
10. Z. Shu, L. Xie, and Q. Li, "Single-phase back-to-back converter for active power balancing, reactive power compensation and harmonic filtering in traction power system," IEEE Trans. Power Electronics, Vol. 26, Issue 2, pp. 334–343, 2011.
11. G. Zhu, J. Chen, and X. Liu, "Compensation for the negative-sequence currents of electric railway based on SVC," in Proc. of ICIEA 2008, 2008, pp. 1958–1963.

12. J. Ma, M. Wu, and S. Yang, "The application of SVC for the power quality control of electric railways," in Proc. of SUPERGEN'09, 2009, pp. 1–4.
13. W. Hosny, H. E. Park and J. H. Song, "Investigation of shunt active power filters in railway systems, substation installation", in Proc. of EPQU 2011, 2011, pp. 1–6.
14. X. Z. Xu, B. C. Chen and F. C. Gan, "Electrical railway active power filter research based on genetic algorithms", proc. of ICCA 2007, 2007, pp. 1465–1468.
15. Y. Horita, N. Morishima, M. Kai, M. Onishi, T. Masui, and M. Noguchi, "Single-phase STATCOM for feeding system of ToKaido Shinkansen," in Proc. of IPEC 2010, 2010, pp. 2165–2170.
16. T. Uzuka, "Faster than a speeding bullet: an overview of Japanese high-speed rail technology and electrification," IEEE Electrification magazine, 2013, pp. 11–20.
17. Z. Shu, S. Xie, K. Lu, Y. Zhao, X. Nan, D. Qiu, F. Zhou, et al., "Digital Detection, control, and distribution system for co-phase traction power supply application," IEEE Trans. Indus. Electro., Vol. 60, No. 5, pp. 1831–1839, May 2013.
18. X. He, Z. Shu, X. Peng, Q. Zhou, Y. Zhou, Q. Zhou, and S. Gao, "Advanced cophase traction power supply system based on trhee-phase to single-phase coverter", IEEE Trans. Power Electro., Vol. 29, No. 10, pp. 5323–5333, Oct. 2014.
19. B. Gholizad, and M. Akhbari, "A topology of hybrid active power filter for simultaneously compensating harmonics and load unbalance in single phase traction systems," in Conf. on Power Engineering, Energy and Electrical Drives, 2011, pp. 1–6.
20. K-W. Lao, N. Dai, W.G. Liu, and M. C. Wong, "Hybrid power quality compensator with minimum DC operation voltage design for high speed traction power systems," IEEE Trans. Power Electronics, Vol. 28, No. 4, pp. 2024–2036, Apr. 2013.
21. N.Y. Dai, K-W. Lao, M-C. Wong, and C-K. Wong, "Hybrid power quality conditioner for co-phase power supply system in electrified railway," IET Power Electronics, Vol. 5, No. 7, pp. 1084–1094, Aug. 2012.
22. K-W. Lao, M-C. Wong, N.Y. Dai, C-K. Wong, and C-S. Lam, "A systematic approach to Railway Hybrid Quality Conditioner design with harmonic compensation," IEEE Trans. Industry Electronics, Vol. 62, No. 2, pp. 930–942, Feb. 2015.
23. N. Y. Dai, M-C. Wong, K-W. Lao, and C-K. Wong, "Modelling and control of a railway power conditioner in co-phase traction power system under partial compensation," IET Power Electronics, Vol. 7, No. 5, pp. 1044–1054, May 2014.
24. Ning Yi Dai; Keng-Weng Lao; Chi-Seng Lam, "Railway Hybrid Quality Conditioner With Partial Compensation for Converter Rating Reduction", *IEEE Transactions on Industry Applications*, 2015, vol: 51, issue: 5, pp. 4130–4138.
25. K. Yu, "Electric railway power supply and power quality," China Electric Power Press, 2010.
26. GB/T 15543-1995, "Quality of electric energy supply admissible three-phase voltage unbalance factor," National Standards of the People's Republic of China.
27. IEEE Std 1159-2009, "IEEE recommended practice for monitoring electric power quality," IEEE, 2009.

Chapter 6
Hardware Construction and Experimental Results

Abstract As presented in the contents of other chapters, experimental results obtained from a laboratory-scaled hardware prototype are used to show the effectiveness of Railway HPQC in co-phase traction power. More details about the hardware prototype construction and experimental results are presented in this chapter. Major contents include hardware schematics, microcontroller control, signal conditioning board, and testing of IGBT drivers.

6.1 Hardware Design and Implementation

Besides simulation, experimental results are used to show the performance of Railway HPQC. Before attempting to obtain experimental results for verification, the hardware is designed and constructed. Details of the circuit schematics, design procedures, as well as component descriptions are given in this section.

6.1.1 Hardware Schematics

The circuit schematic of the hardware prototype is shown in Fig. 6.1.

The system is constructed with a ratio of 1:550 downscaled from actual locomotive load voltage of 27.5 kV. System descriptions are shown below.

- Rated operation load voltage RMS: 50 V.
- Rated load current RMS: 3 A.
- Full load rated capacity: 150 VA.

The hardware prototype is constructed at 50 V rms load voltage level, with rated full load capacity of 150 VA. The voltage level is reduced by a factor of 550 compared to simulation verifications in previous contents. The V/V transformer is composed of two single-phase transformers (220/220 V and 220/110 V), each of

K.-W. Lao et al., *Co-phase Traction Power Supply with Railway Hybrid Power Quality Conditioner*, https://doi.org/10.1007/978-981-13-0438-5_6

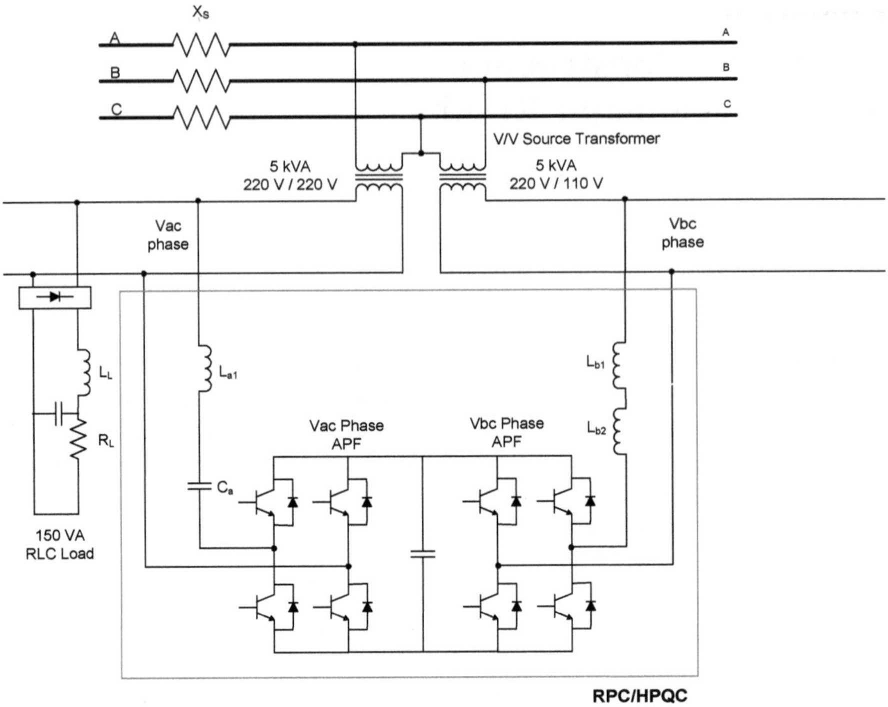

Fig. 6.1 Circuit schematic of the co-phase traction power with HPQC hardware prototype

5 kVA capacity. The Railway HPQC parameters are then designed according to the minimum operation voltage requirement (Chap. 3). Further details can be found in later sections.

With Fig. 6.1 alone is still not enough for hardware construction, the linkage of circuit components with real circuit components should also be provided. Figure 6.1 is then revised as hardware schematic showing all physical hardware components, as shown in Fig. 6.2. Notice that besides main power components, other gadgets like connection terminals, transducers, and operation panel components are also shown in the figure. The load circuit schematic for the traction load model is also shown in Fig. 6.3. It is being modeled by a rectifier RL load and is connected across the Vac phase terminal.

The power supply structure for signal condition circuits and other gadgets of hardware prototype is shown in Fig. 6.4.

6.1.2 Microcontroller

The microcontroller used in the hardware prototype is TDS2812EVMB (DSP2812) from Wintech. This model is chosen since other gadgets are available for evaluation, and that there are internal PWM generator, comparator, and timer as well as

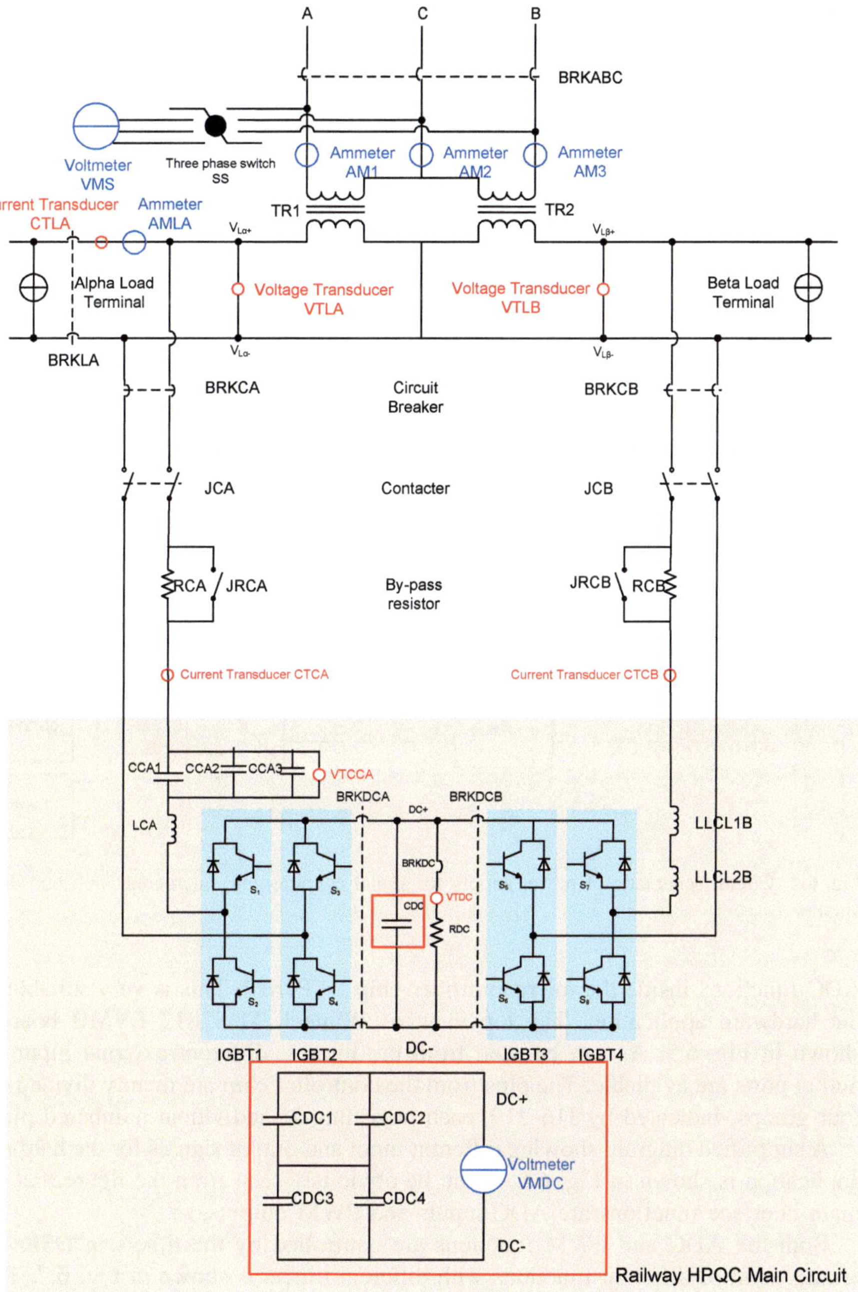

Fig. 6.2 Hardware circuit schematics of co-phase traction power supply prototype with Railway HPQC

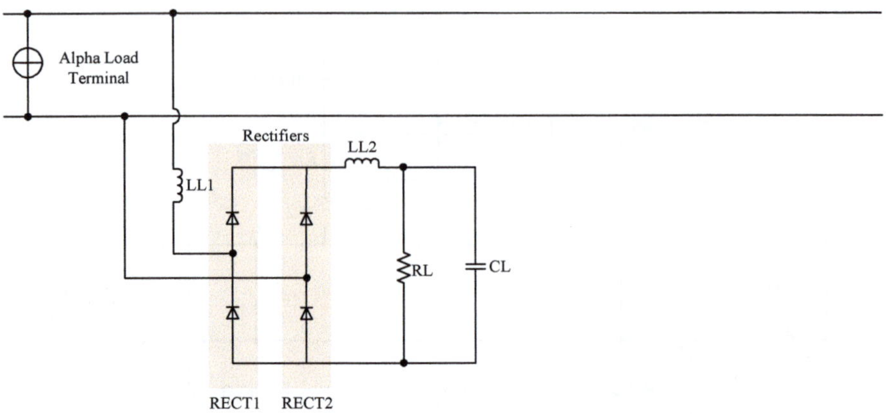

Fig. 6.3 Circuit schematics of rectifier RLC load model for traction load

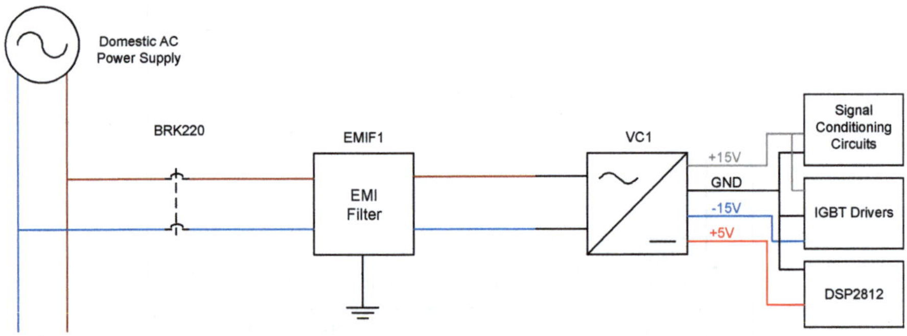

Fig. 6.4 Circuit schematic of power supply for signal conditioning, microcontroller, and driver circuits

ADC functions inside the microcontroller chip DSP2812. This is very suitable for the hardware application. The top view of Wintech TDS2812 EVMB board is shown in Fig. 6.5. As can be seen from the figures, different external input and output ports are available. The pins from the controller chip are mainly divided into four groups, indicated by J16–J19, each providing 36 individual numbered pins.

A simplified diagram showing different input and output signals for the hardware application is shown in Fig. 6.6. It can be obviously seen from the figure that two main interface functions are ADC inputs and PWM outputs.

Both the ADC and PWM functions are controlled by the timers in DSP. The arrangement of different functions with different timers is shown in Fig. 6.7. They are briefly discussed below.

Fig. 6.5 Top view of Wintech TDS2812EVMB board

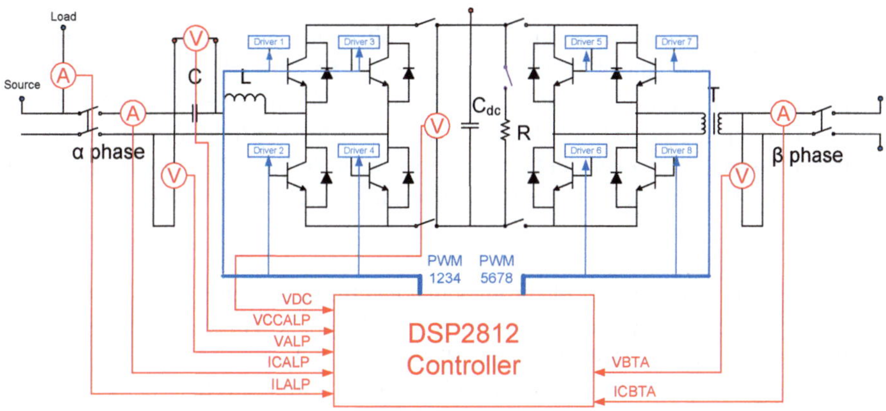

Fig. 6.6 Simplified diagram showing the required input and output signals for microcontroller in the hardware application

ADC Inputs

Since DSP2812 is a digital signal processor, all processing is done digitally. Analog/Digital Conversion (ADC) is thus required to transform the analog signals in the real world to digital signals for computation. There are totally 14 12-bit ADC inputs (ADCINA0 ~7 and ADCB0 ~7), which may work in cascade or series mode in DSP2812. As shown in Fig. 6.7, six ADC input pins are being used in the

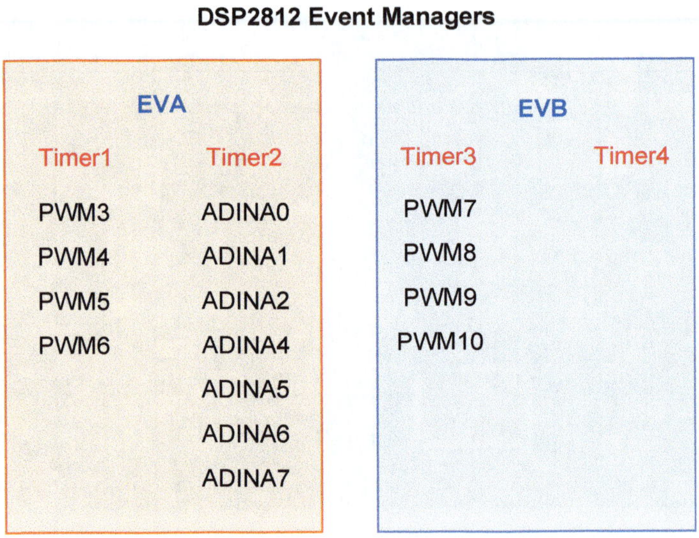

Fig. 6.7 Simplified diagram showing the arrangement of different timers with different functions

hardware prototype. There is an internal 12-bit ADC module, and the converted results are stored in the ADCRESULT register. Besides, the sequence and frequency of ADC may be selected by defining different ADC control registers. The ADC sampling period may be controlled by either the timer or by defining the control registers. However, details will be neglected in this report. For more information, the user menu from Texas Instrument may be referred.

PWM Outputs

It is well known in power electronics that electronic switches (such as the IGBT in this hardware application) are driven by PWM signals. There are totally six pairs of PWM output signals (PWM1–6 and PWM7–12), which can be generated by the compare register. PWM1–6 is controlled by timer1 while PWM7–12 is controlled by timer3. The PWM signals are generated using the internal comparator in DSP2812, and they may be chosen as active low or active high by defining PWM control registers. In addition, it is also essential in power electronics that dead band should be provided in PWM signal pairs so as to avoid short-circuit conditions. There are also internal dead band generators for PWM signals, whose dead band period can also be tuned using DBTCON registers.

The connections of ADC input and output signals with other electronic gadgets are shown in Fig. 6.8. Details of transducers and IGBT drivers will be given in later sections.

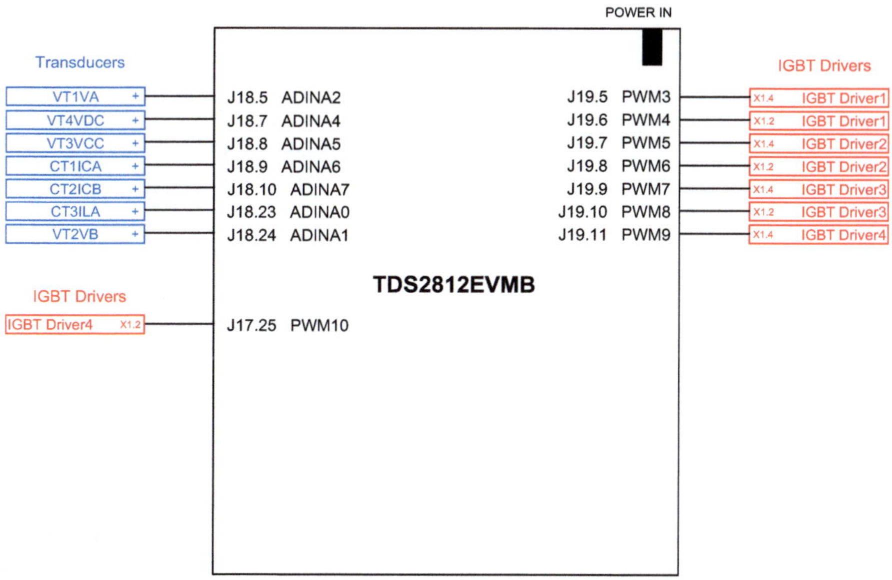

Fig. 6.8 Connections of TDS2812EVMB ADC and PWM pins with other electronic gadgets

6.1.3 Signal Conditioning Circuits

As discussed above, ADC is required to transform physical signals into digital ones. However, the physical signals from the hardware are too large and may not be capable for direct DSP input. Signal conditioning is thus required. Transducers are used to transform large physical signals into electrical ones. The two transducers used in the hardware are KV50A/P (voltage transducer) and KT20A/P (current transducer). The outputs from the transducers may still not be capable for DSP input and further processing is required. For instance, in TDS2812EVMB, the ADC input signal should be within 0–3.3 V.

Shown in Fig. 6.9 is the circuit schematic of the signal conditioning circuit. It is mainly divided into three stages.

In order to synthesize the parameters in the signal conditioning circuit, the value of R1 in Fig. 6.9 is chosen as 100 O for KV50A/P (voltage transducer) and 50 O for KT20A/P (current transducer). This ensures that the input signal is of 5 Vrms value.

Other circuit parameters may be selected as follows. The input signal v_{in} is to be transformed into one that is capable for DSP. The extreme input signal amplitude is thus transformed into a range of (0.3–3.3 V). The resistance values R2, R3, and R5 may then be designed according to (6.1). Notice that the peak of v_{in} is $5\sqrt{2}$V instead of 5 V.

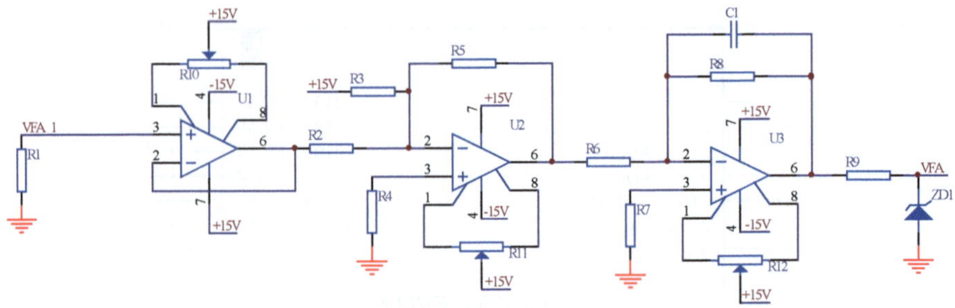

Fig. 6.9 Circuit schematic of the signal conditioning circuit used in the hardware prototype

$$v_o = \frac{R5}{R2} v_{\text{in}} + \frac{R5}{R3} \times 15 \tag{6.1}$$

For example, the following parameters may be selected. R2 = 10 kOhm; R3 = 18 kOhm; R4 = 1.5 kOhm; R5 = 2 kOhm; R6 = 10 kOhm; R7 = 5.1 kOhm; and R8 = 10 kOhm. R10, R11, and R12 are all 20 kOhm for zero adjustment.

The appearance of the signal conditioning circuit in the hardware prototype is shown in Fig. 6.10.

6.1.4 IGBT Drivers

As discussed above, the electronic switches are driven by PWM signals. IGBT drivers are thus required to provide electrically isolate signals to various IGBTs located within different circuit levels. The model of the IGBT driver used in the hardware is POWERSEM PSHI23. The appearance of the IGBT driver is shown in Fig. 6.11.

Preliminary testing is done before installing the driver into the hardware. Some of the pins need to be short-circuited in order to choose the appropriate logic level and to output the entire PWM signal. The pin connections are shown in Fig. 6.12.

PWM signals of different duty ratios and various frequencies are being tested. The inputs and outputs are monitored. Two typical switching frequencies 1 and 10 kHz are being tested. The tested results are shown in Table 5.1. It can be observed from the statistics that the output signal duty ratio follows that of the input one.

For reference, the input and output waveforms of one typical case (duty ratio = 0.5) for each case (1 and 10 kHz) are presented in Fig. 6.13. It is found that the turn-on delay is 2 us while the turn-off delay is approximately 1.6 us.

Fig. 6.10 Appearance of the signal conditioning circuit in hardware prototype

6.1.5 Hardware Appearance

The hardware is being implemented and constructed using a webcase. The external appearance of the hardware prototype is shown in Fig. 6.14. The layout design is also shown in Fig. 6.15. It is mainly divided into four layers. With this design, the control circuits are located far away from the main power components to reduce EMI effect. Details of the arrangement inside will be explained below.

Fig. 6.11 POWERSEM PSHI23 IGBT driver

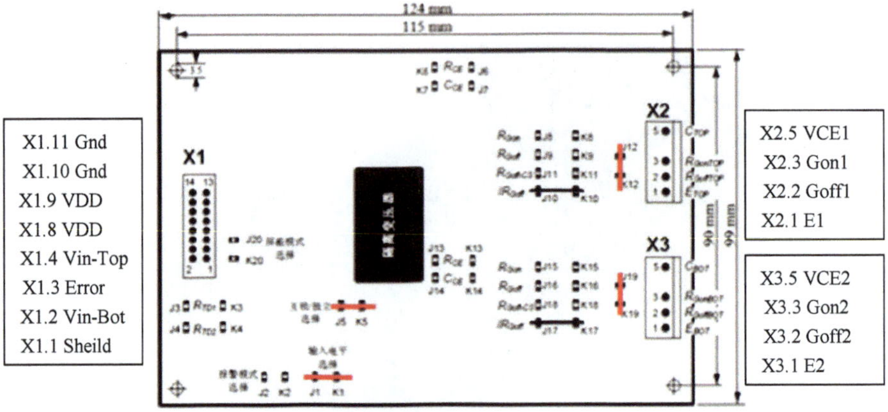

Fig. 6.12 Pin connections of the IGBT driver PSHI23

Table 6.1 Tested results of IGBT driver with different duty ratios (1 and 10 kHz)

Input frequency (kHz)	Input duty ratio (%)	Output frequency (kHz)	Output duty ratio (%)
1.119	20.4	1.119	20.4
1.101	41	1.101	41
1.096	60.5	1.096	60.5
1.116	79.4	1.119	79.4
11.24	20.3	11.24	20.3
11.11	40.2	11.11	39.8
11.01	60.0	11.01	59.7
11.26	79.5	11.26	79.2

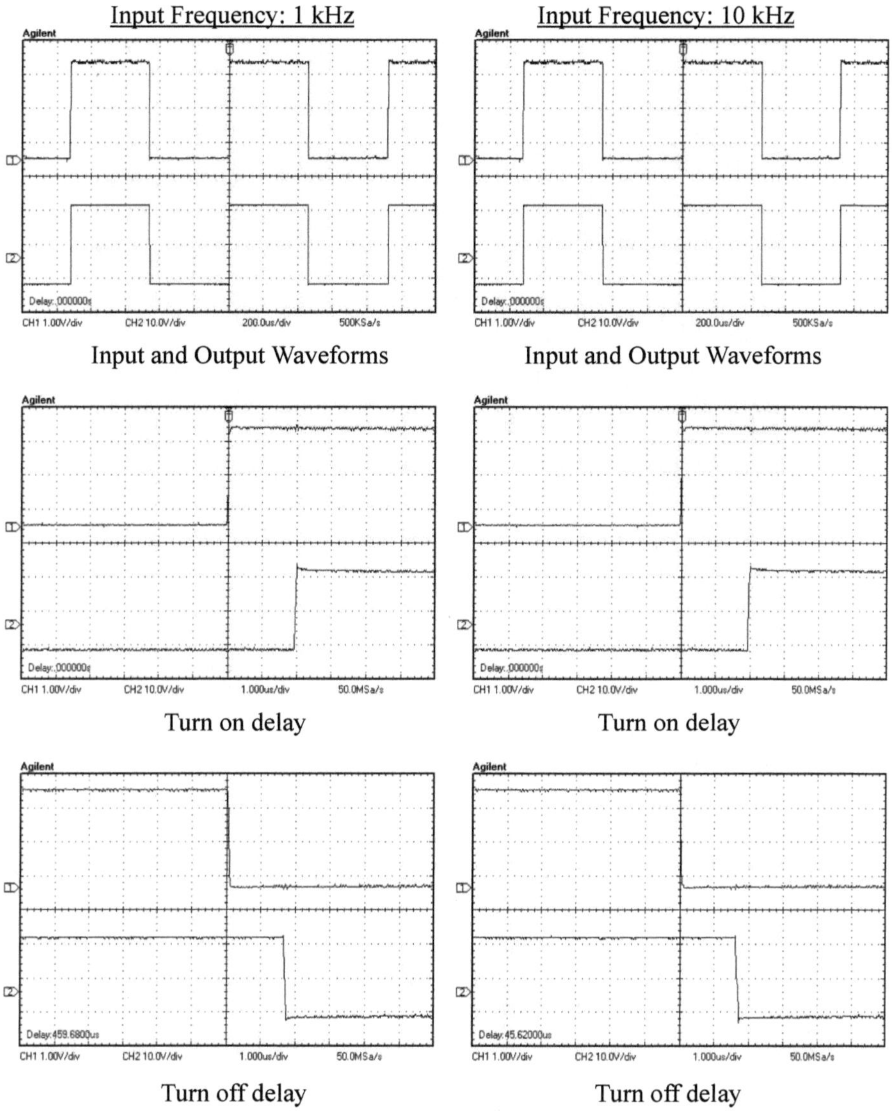

Fig. 6.13 Input and output waveforms obtained during testing of IGBT driver

Front View (Operation Panel)

The appearance of hardware prototype of co-phase power with Railway HPQC and the testing area is given in Fig. 6.16. The front view of the hardware prototype mainly includes the operation panel, as shown in Fig. 6.17. All the voltmeters and ammeters, control buttons, LED indicators, as well as circuit breakers are located at

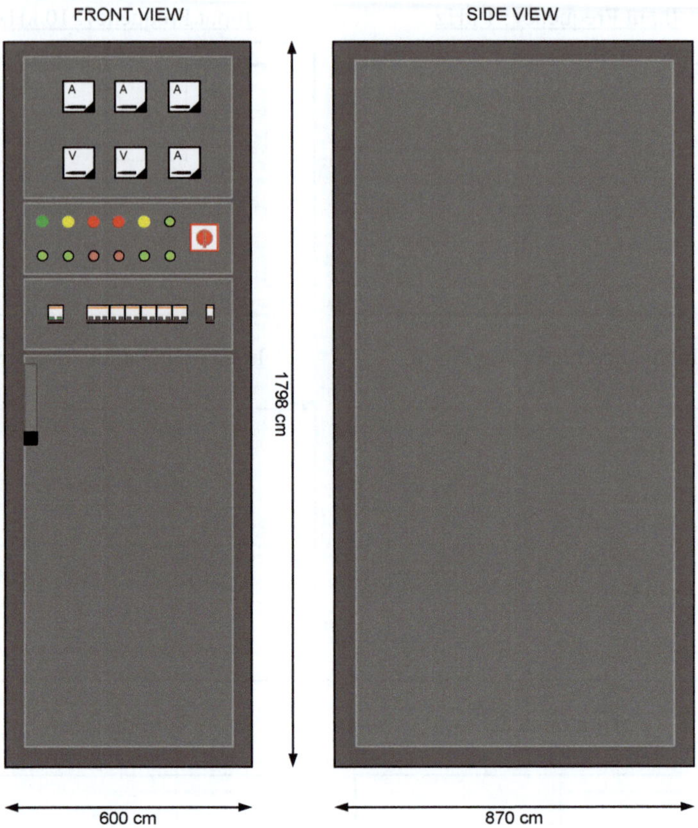

Fig. 6.14 External appearance design of the hardware prototype

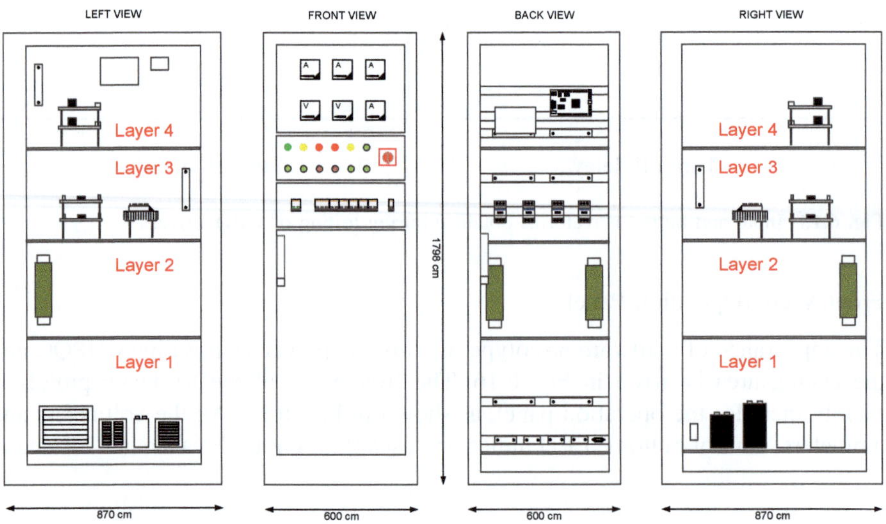

Fig. 6.15 Layout design of the hardware prototype

Fig. 6.16 Appearance of hardware prototype of co-phase power with Railway HPQC (laboratory-scaled)

the front side. This layout is designed so as to ease the operation of the operators in controlling and monitoring the hardware performance during experiments. Details of the operation functions may be found in Appendix A.

Back View (Control and Connect Terminals)

The back view of the hardware prototype can be found in Fig. 6.18. It mainly includes the DSP controller, input terminals, as well as circuit contactors. As will be described in later sections, the power components are placed at the first layer, and the microcontroller is thus located at the fourth layer to reduce the EMI effect. Besides, all terminals are named to make it easier for debugging.

Besides the front and back views, the whole hardware prototype is divided into four layers, with first layer as the bottom and fourth layer as the top. Descriptions of each layer can be found below.

First Layer (Power Components)

The appearance of the first layer is shown in Fig. 6.19. All power components are placed at this layer since they are usually heavy in mass. They are better to be located at the bottom layer; otherwise, the effort for the hardware stand is heavy.

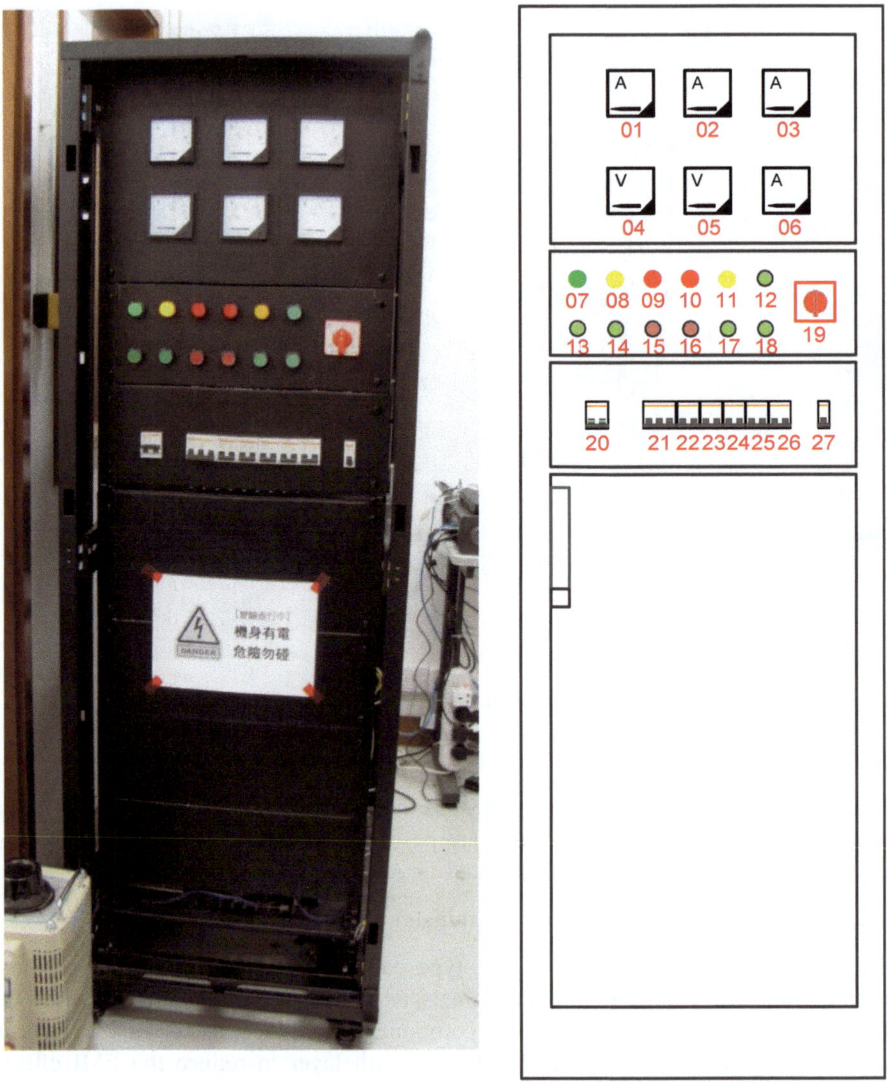

Fig. 6.17 Front view of the hardware prototype

Second Layer (Reserved)

The appearance of the second layer is shown in Fig. 6.20. This layer is mainly reserved for further development and only two bypass resistors are located here.

Third Layer (IGBT and IGBT Drivers)

The appearance of hardware third layer can be found in Fig. 6.21. They mainly include the IGBTs and IGBT driver. These components are located in this layer to

Fig. 6.18 Back view of the hardware prototype

avoid the EMI effect caused by the switching affecting the performance of the DSP located in the fourth layer. In addition, the distance between the IGBT and its driver can be minimized to avoid undesired PWM signal distortions affecting their performances.

Fourth Layer (Signal Conditioning Circuit)

The appearance of the hardware fourth layer can be found in Fig. 6.22. Besides the microcontroller, the signal conditioning circuits are located in this layer. In addition, the voltage converter is also placed here.

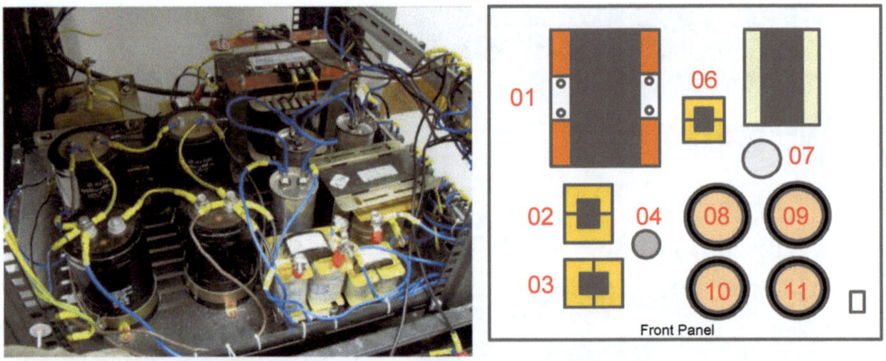

Fig. 6.19 Appearance of hardware first layer

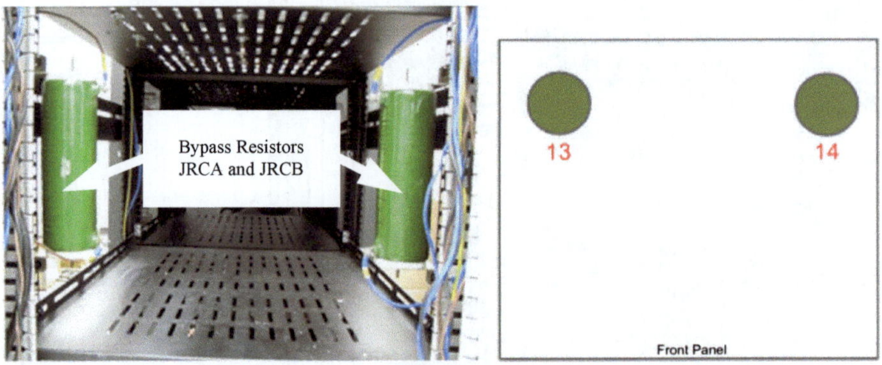

Fig. 6.20 Appearance of hardware second layer

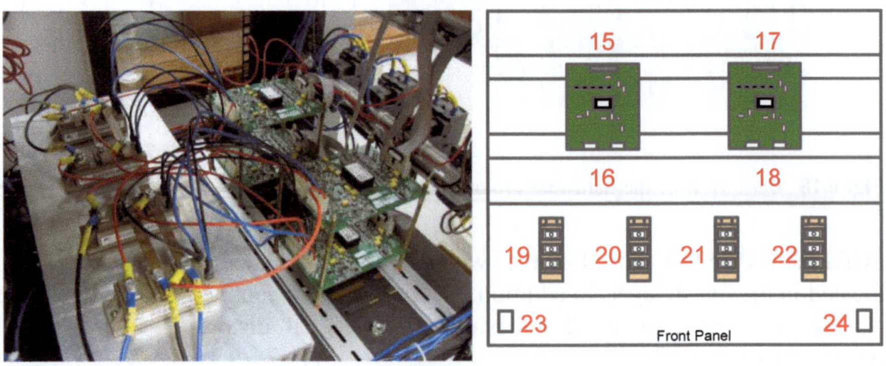

Fig. 6.21 Appearance of hardware third layer

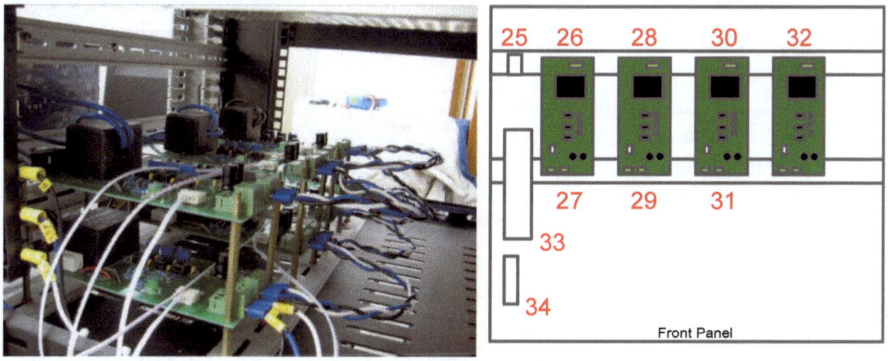

Fig. 6.22 Appearance of hardware fourth layer

6.2 Control Algorithm

The control block diagram used in the hardware prototype is shown in Fig. 6.23. The control is mainly divided into two parts: the Vac phase and Vbc phase computations. The load instantaneous real and imaginary power are first computed; afterwards, the required Vac and Vbc phase compensation active and reactive power can be determined. The reference compensation current is thus generated and input to a hysteresis PWM controller, generating required PWM signals to the IGBT switches.

The control flowchart may be found in Fig. 6.24. The whole program mainly includes six main processes. After system, ADC, and PWM initializations, ADC signals are input and computations are done whenever ADC Interrupt (ADCINT) is encountered. PWM signals are also refreshed during computations. Afterwards, the ADC interrupt will be reset to prepare for another ADC trigger interrupt.

The settings of different functions are summarized in Table 6.2.

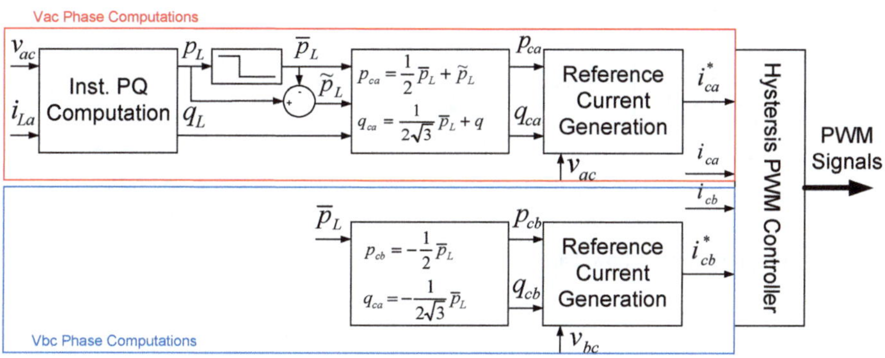

Fig. 6.23 Control block diagram showing the control in hardware application

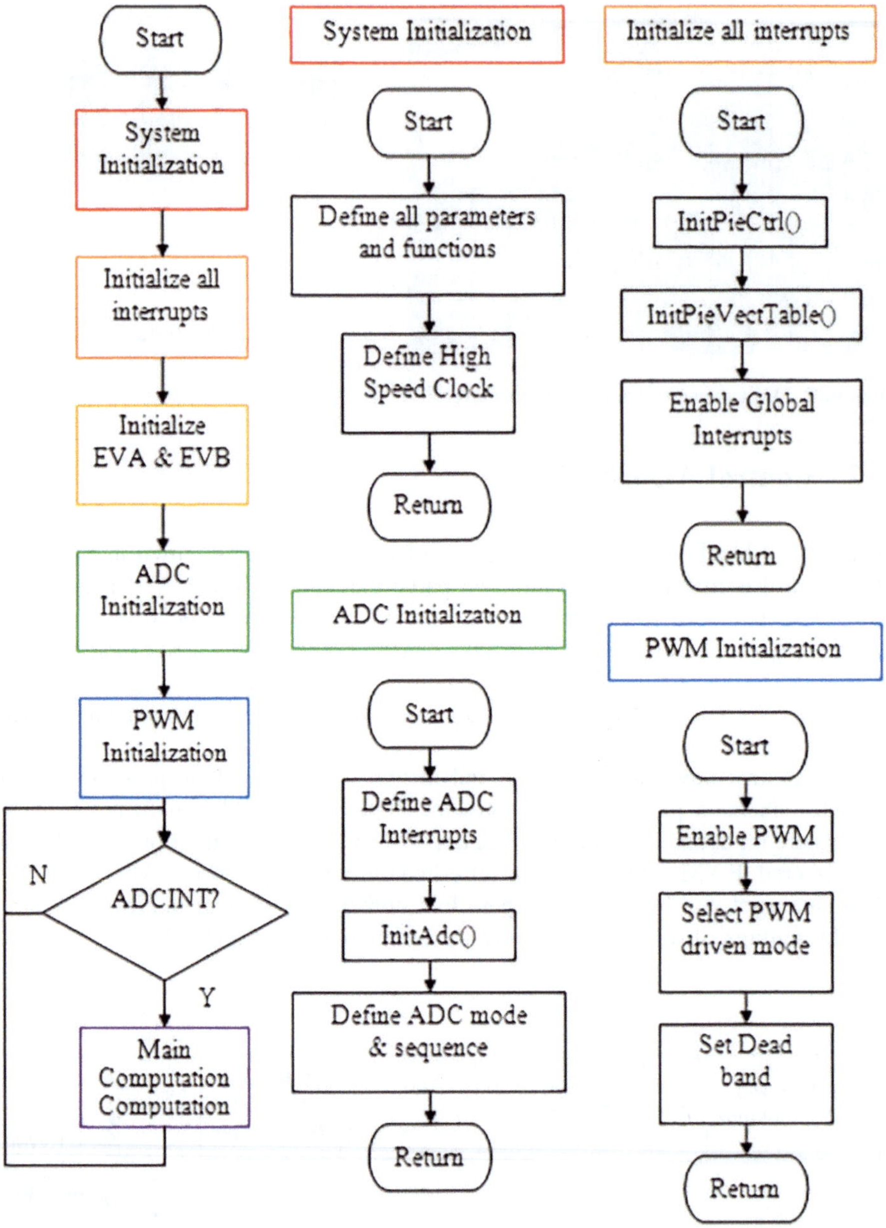

Fig. 6.24 Program control flowchart

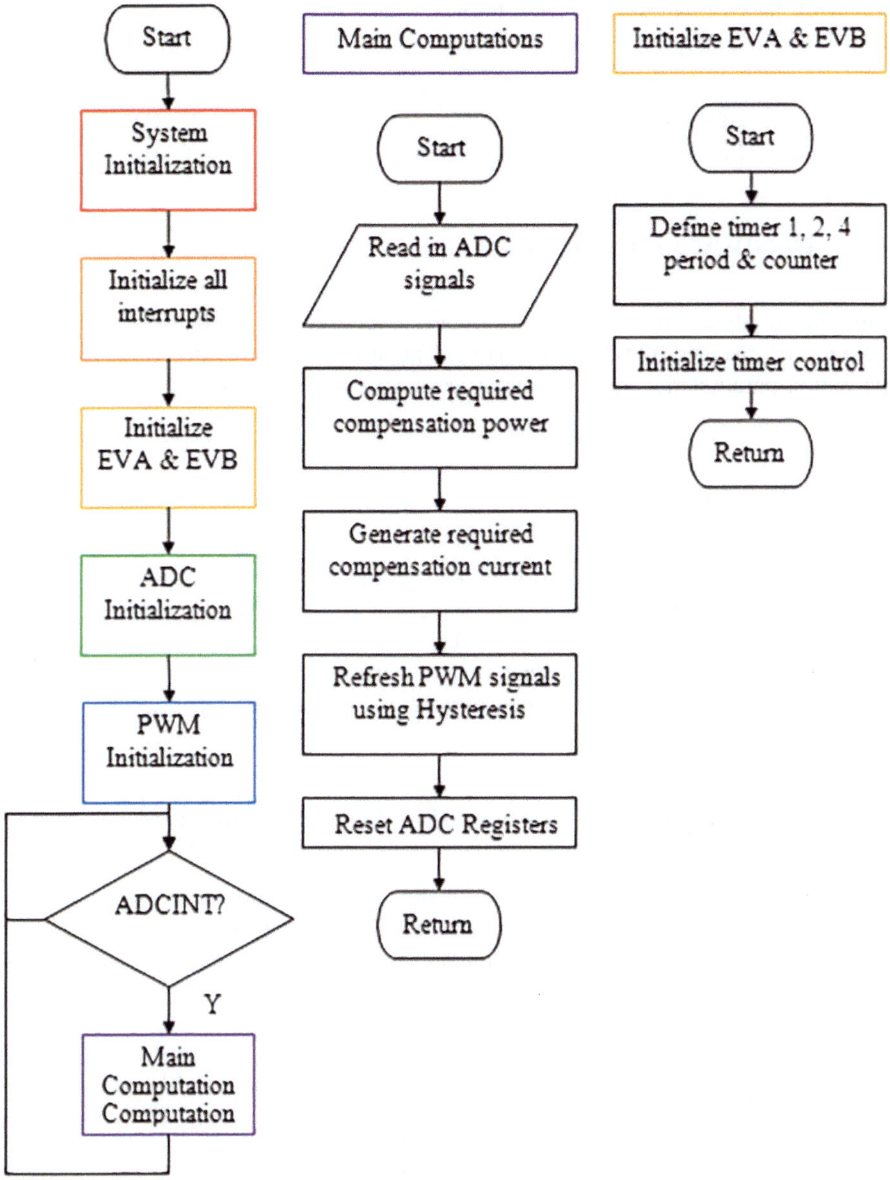

Fig. 6.24 (continued)

Table 6.2 Summarized data of different function settings in the experiment

No.	Item(s)	Description(s)
1	System clock frequency (after setting HISPCP)	37.5 MHz
2	ADC interrupt frequency	20 kHz
3	Maximum PWM switching frequency	20 kHz

6.3 Hardware Parameters

Different hardware parameters used in the experimental verifications are shown below. For instance, they include (i) load parameters and (ii) power quality compensation parameters. They are shown here for comparison and reference.

6.3.1 *Load Parameters*

Throughout the whole project, different experiments are done to verify the theory and design. Most experiments in Chap. 3 are performed based on rated load condition with rated load capacity ($r = 1$), while in Chap. 4, load variations are introduced into the experiments that the load capacity changes from 0.2 to 1.6. With reference to Fig. 6.3, details of load parameters are shown in Table 6.3.

6.3.2 *Power Quality Compensation Device Parameters*

In this project, two different power quality compensation devices in co-phase traction power supply, namely, conventional RPC and new Railway HPQC, are investigated in the experimental verifications. The circuit parameters used for the operation under rated load condition in Chap. 3 are shown in Table 6.4.

Table 6.3 Details of the load parameters in the experiments of this research project

Load capacity(r)	LL1 (mH)	LL2 (mH)	RL (ohm)	CL (uF)
0.2	36.3	40.5	103	50
0.4	19.5	36.9	55	70
0.6	9.0	5.1	34	170
0.8	9.0	5.1	26	220
1.0 (rated)	9.0	10.5	21	260
1.2	2.0	10.5	17	260
1.4	2.0	10.5	14	220
1.6	2.0	10.5	10	220

Table 6.4 Detailed circuit parameters of the experimental verification under fixed rated load condition in Chap. 3

			RPC	New Railway HPQC	
			Traditional design	Traditional hybrid filter design	Filter design
No.	Parameters		Value	Value	Value
1.	Vac phase coupled impedance ratio	m_{La}	0.074	0.876	0.876
2.	Vac phase coupled inductance ratio	k_L	NA	0.042	0.10
3.	Vac phase coupled inductance	L_a	4 mH	2 mH	4.9 mH
4.	Vac phase coupled capacitance	C_a	NA	170 uF	150 uF
5.	Vbc phase coupled inductance	L_b	4 mH	4 mH	4 mH
6.	dc link capacitance	C_{dc}	10,000 uF	10,000 uF	10,000 uF
7.	RPC operation voltage rating	k_{invaL}	1.13	0.5131	0.54
8.	dc link operation voltage	V_{dc}	80 V	40 V	38 V

It can therefore be observed from the table and the contents in Chap. 3 that the performance of conventional RPC and Railway HPQC in providing power quality compensation in co-phase traction power supply is similar. However, the operation voltage of novel Railway HPQC using new filter design is only around 50% of traditional RPC. This can lead to 50% reduction in operation loss and initial cost.

6.4 Summary

In this chapter, details of the laboratory-scaled hardware prototype of co-phase traction power supply with Railway HPQC are discussed. For safety concern, the operation voltage is scaled down to 1:550. The hardware schematic is designed based on the simulation schematics during verifications, with rectifier RLC load as traction load model. The whole control is accomplished using Digital Signal Processor (DSP) 2812 (TDS2812 EVMB board). The power parameters such as current and voltage are transformed into electrical signals that are suitable for DSP programming and processing using signal conditioning circuit. The power electronics switches used are IGBTs, and they are controlled by the PWM signals generated by the DSP. IGBT drivers are used to drive the IGBTs, and the testing results of the IGBT driver can be found within this chapter. Details of the hardware parameters can also be checked within this chapter.

Chapter 7
Summary

Abstract This chapter summarizes the aspects related to co-phase traction with Railway HPQC. The summary includes review of major problems and challenges of high-speed railway, benefits of co-phase traction power supply structure as well as potential direction on relevant research and development.

7.1 Major Problems and Challenges in High-Speed Railway Traction Power Supply System

7.1.1 High-Speed Requirement and Essential Need of New Topology

High-speed achievement is important for high-speed railway. In conventional or existing traction power supply system, electric power is supplied from two different phases. Since locomotives all move along the same contact wire headline, insulators (isolation components) are required and installed between substations. When locomotives run through areas with isolation components, they lose velocity and high-speed cannot be achieved. On the other hand, in newly developed co-phase traction power, electric power is supplied to locomotives so that they can run along the same contact wire headline. It has already been proved that co-phase traction power supply is advantageous for higher power supply reliability, transformer utilization ratio, and comprehensive power quality compensation management.

7.1.2 Low Short-Circuit Capacity and Ability to Withstand System Unbalance

Short-circuit capacity is the apparent power that the system will dissipate when the system is short-circuited. The value of the short-circuit capacity is actually inversely proportional to the system source impedance. In other words, a lower short-circuit

K.-W. Lao et al., *Co-phase Traction Power Supply with Railway Hybrid Power Quality Conditioner*, https://doi.org/10.1007/978-981-13-0438-5_7

capacity indicates a higher source impedance. In some traction power supplies, the short-circuit capacity is low such that the system cannot withstand high system unbalance. However, it is rare that locomotive loadings are balanced since it depends on individual transportation demand. This makes system unbalance a problem in traction power supply with low short-circuit capacity and must be solved.

7.1.3 Inductive Locomotive Loadings and Low Power Factor

Since locomotive loadings are mostly inductive, they cause the presence of reactive power in the system source. Presence of reactive power system indicates inefficient usage of the power factor and can be evaluated by power factor (defined as the ratio of active power to apparent power). When the value of power factor is less than 1.0, reactive power is present and can threaten system stability.

7.1.4 Usage of Power Electronics in Locomotive Loadings and High Harmonic Distortions

With the exclusive usage of nonlinear power electronic components in locomotive loadings, harmonic distorted current is injected into the system and causes high harmonic distortions. Presence of harmonics in the system not only indicates inefficient usage of energy but will also generate heat and will reduce the lifetime of the circuit components. Therefore, actions need to be taken to solve the harmonic problems.

7.2 Railway HPQC Can Act as Support for Development of Co-phase Traction Power Supply Investigation

There is a worldwide trend in developing high-speed railway system for city and country development. It is therefore important to develop a safe, efficient, and high-speed railway system.

As introduced, compared to conventional system which locomotive power is supplied via two different phases, newly developed co-phase traction power in which locomotive power is supplied via one single phase only possesses advantages such as higher power supply reliability and transformer utilization ratio. Co-phase traction power supply system is investigated and developed by the research team in Southwest Jiaotong University and is one of the important projects supported by the Chinese government. In fact, the world's first co-phase traction device has already been put into trial operation in Kunming Meishan substation. Recently, China is having the world's leading role in high-speed railway development.

The research project is an important extension of the investigation of the government supported "co-phase traction power" project. With the investigations and design techniques, the operation voltage, operation loss, and initial cost of the system can be reduced. This can help to enhance the development and implementation of co-phase traction power supply and make contribution to high-speed railway in China.

7.3 Advantages of Co-phase Traction Power System with Railway HPQC

The high-speed traction power system in this book is composed of co-phase traction power with Railway HPQC. Compared to existing co-phase traction power with conventional RPC, the main advantages of co-phase traction power system with Railway HPQC include the following:

- Reduction of operation voltage
- Lower device ratings
- Reduced operation loss
- Lower initial cost

The advantages listed above are mainly by adopting hybrid structure in the coupled impedance. The uniqueness and difficulties in the design are discussed within this report.

The co-phase traction power with Railway HPQC is beneficial for the development of high-speed railway. In co-phase traction power, the number of neutral sections can be eliminated so that locomotives can achieve high speed. Moreover, power quality compensation can be provided by Railway HPQC with low dc link operation voltage to reduce the amount of reactive power and inefficient usage of the energy in power source grid. Power quality enhancement also helps to increase system stability and reliability. This helps to form a safe, efficient, and high-speed railway.

7.4 Analysis of Railway HPQC Operation in Co-phase Traction Power and Its Uniqueness

In order to investigate the co-phase traction power supply system with Railway HPQC, the system and power quality problems are modeled. It is found that during power quality compensation, active and reactive power transfer is required to be accomplished by the compensation device, for instance, Railway HPQC in the system. As hybrid LC structure is adopted in the system, involvement of active power during compensation makes its design unique and different from traditional hybrid filter. In this report, the Railway HPQC operation in co-phase traction power supply is deeply analyzed and the design techniques under different conditions are described and investigated. Detailed analysis can be found in Chap. 2.

7.5 Different Recommended Design of Railway HPQC According to Various Conditions

7.5.1 Design Under Fixed Rated Load Condition

First of all, the design of co-phase traction power with Railway HPQC is developed and investigated under fixed rated load condition. It is found through phase diagram and mathematical analysis that the operation voltage can be minimized when the output inverter voltage of Railway HPQC is in-phase with the output compensation current. Furthermore, a new harmonic filter parameter design is developed to further reduce the operation voltage when considering harmonic compensation. Unlike traditional design which harmonic filter is tuned to the harmonic frequency which load harmonics are mostly dominant at, the developed harmonic filter design is developed based on the harmonic distribution. It is shown through simulation and experimental results that compared to conventional system with RPC, the operation voltage of Railway HPQC can be reduced by around 55%. This will also lead to a decrease of half in the initial cost.

7.5.2 Design Within Load Condition Variation Range

Under some conditions, the load condition may vary and some actions need to be taken when using co-phase traction power with Railway HPQC. They are also being discussed and investigated in this project. As load varies, the required compensation active and reactive power also changes. The unsatisfactory performance during load variations is mainly caused by the insufficient compensation capability.

Increased operation voltage

The most straightforward method to increase the Railway HPQC compensation capability is to increase the operation voltage. The relationship between operation voltage and compensation capability is discussed and investigated. The variations in the required compensation power are also modeled. It can be shown through simulation and experimental results that the system performance is satisfactory within load variation range. This method is good that no hardware changes are required. However, this method requires high operation voltage (although still lower than that of conventional RPC).

Impedance-mapping technique

Although the operation voltage of Railway HPQC can be increased to increase the compensation capability, this will increase the device rating and operation loss. Moreover, the coupled capacitance is high and can induce high cost. Therefore, another method is developed for reduction of operation voltage and coupled capacitance. The method works by mapping the coupled impedance with the load variation range. The relationship between them is first analyzed and detailed design procedure is developed. The criteria which reduction in operation voltage can be achieved are also presented. Other methods are preferred when the criteria are not satisfied. It is shown through simulation and experimental results that there is around 10% reduction in operation voltage and 30% reduction in coupled capacitance compared to the method of increased operation voltage during load variation range. However, modifications in the coupled impedance are required in this method and sometimes changes may not be possible once installed.

Adaptive dc link voltage control technique

As mentioned previously, under some circumstances when the criteria for reduction of operation voltage and coupled impedance are not satisfied other methods are prepared. However, in the first method when the Railway HPQC operation voltage is increased, the operation loss is also increased. Therefore, the adaptive dc link voltage control technique is investigated. Unlike the first method, the operation voltage of Railway HPQC is automatically adjusted according to the load condition so that the operation voltage can be at an extent minimized during load variations. The theory is also verified via simulation and experimental results.

Details of the above design procedure and analysis of mentioned techniques for load variation condition can be found in Chap. 4.

7.6 Further Potential Development

With China as one of the world's leading role in high-speed railway development, investigations of Railway HPQC can help to further enhance the development of high-speed railway. The research project can make great contributions and worth further investigation. Some further analysis and development of the project are listed below.

7.6.1 Investigation of Multilevel Structure

Although the operation voltage of Railway HPQC is effectively reduced in the system and can be used in multilevel structure, for further development and implementation of Railway HPQC in Railway HPQC, it is better to develop

Railway HPQC with multilevel topology that is suitable for control and power transfer between different phases in traction power. Major challenges in the multilevel topology include MMC technologies and control strategies.

7.6.2 *Exploration of Other Possible Coupled Impedance Structure*

In this project, a hybrid structure is adopted for the coupled impedance in Railway HPQC. The main objective is to reduce the operation voltage and other parameter designs are not discussed as long as the system performance is satisfactory. Actually, other coupled impedance techniques such as LCL filter can be used and it is worth comparing the harmonic suppression performance and operation voltage requirement between Railway HPQC using hybrid LC branch and LCL filters. This can further enhance the performance of the system.

7.6.3 *Transition Actions Between Changes from Conventional RPC to the New Railway HPQC System*

In fact, the topology of co-phase traction power supply is somehow different from the existing one. Therefore, certain techniques are required to transform existing traction power supply into co-phase one with Railway HPQC. Besides, it would be beneficial if the usage of Railway HPQC can be extended to existing traction power supply in which locomotive power is obtained from two different phases. This can help to minimize the effect between the transitions from conventional system to newly co-phase traction power with Railway HPQC.

7.6.4 *Extension of Co-phase Power with Railway HPQC Technique to Smart Grid*

Besides application in traction power, the idea of co-phase power with Railway HPQC can also be extended into smart grid. As three-phase ac and single-phase ac and dc subsystems are also present within the circuit topology, it can be used in ac and dc hybrid microgrid and can provide power quality compensation at the same time. In other words, Railway HPQC idea can be used to develop an energy management device that can be plugged and played and can be connected between ac and dc microgrid to provide power and power quality management.

7.7 Final Remarks

With the trial operation of the world's first co-phase device in the MeiShan railway substation, and the development in the research of co-phase traction power, the co-phase traction power supply has high potential to have large scale implementation soon. Its impact to the power system is one of the major concerns. Therefore, the authors believe that in the following years there will be continuous research on high speed traction railway such as co-phase traction power, its power quality conditioning and management techniques, as well as the implementation of the system.

MIX
Papier aus verantwortungsvollen Quellen
Paper from responsible sources
FSC® C105338

If you have any concerns about our products,
you can contact us on
ProductSafety@springernature.com

In case Publisher is established outside the EU,
the EU authorized representative is:
Springer Nature Customer Service Center GmbH
Europaplatz 3, 69115 Heidelberg, Germany

Printed by Libri Plureos GmbH
in Hamburg, Germany